iPad Air (4th Generation) User Guide

The Complete Illustrated, Practical Guide with Tips & Tricks to Maximizing the latest iPad Air 4th Generation

Daniel Smith

Content

Introduction

Apple has introduced an all-new iPad Air — the most powerful, versatile, and colorful iPad Air ever. Now available in five gorgeous finishes, iPad Air features an all-screen design with a larger 10.9-inch Liquid Retina display, camera and audio upgrades, a new integrated Touch ID sensor in the top button, and the powerful A14 Bionic for a massive boost in performance, making this by far the most powerful and capable iPad Air ever made.

The new iPad Air 4 (2020) marks a big change for Apple's 'light-as-air' line of tablets - no longer is it an ungainly version of the 'standard' iPads, but it's now more like a specced-down iPad Pro.

This 2020 model in the line, the iPad Air 4, got shown off at Apple's September event alongside the entry-level iPad (2020), the Apple Watch 6 and Apple Watch SE. It was certainly the most premium product shown off at the event, and maybe the most intriguing too.

This is the fourth generation of Apple's mid-range 'iPad Air' line, and it's a family Apple didn't seem to know what to do with, as the iPad Air 3 was just a bigger version of the entry-level iPad in many ways. The iPad Air 4 has loads in common with the iPad Pro (2020) models, so it seems Apple has a plan for the Airs.

To help you get your head around this tablet, we've compiled all the information we have on it so far including the iPad Air 4's price, release date, specs and features.

iPad Air (2020) release date and price

The new iPad Air price starts at $599 / £579 / AU$899, and it will be available to buy from 'next month', meaning some time in October 2020. That's quite a hike over its predecessor, but it seems like we're getting a big specs jump too.

That's the starting price, but you can get it in both 64GB and 256GB, with LTE connectivity or just Wi-Fi if you want. We'll put a graph below with all the prices.

The iPad Air 3 started at $499 / £479 / AU$779 for a version with 64GB storage, and $649 / £629 / AU$999 for 256GB. The price went up $130 / £120 / AU$200 for each for the LTE version, instead of Wi-Fi.

iPad Air design and display

The iPad Air has a flat edge, an all-screen display, a single rear camera and some subtle edge buttons – it looks a lot like an iPad Pro, which is a change from previous iPad Airs, which looked like large versions of the entry-level slates with big bezels and physical front buttons.

The screen uses Apples Liquid Retina display tech, which is a fancy form of LCD. It has a resolution of 2360 x 1640 and it's 10.9 inches across, which isn't quite as big as the iPad Pro models, as the smallest size of that slate is 11 inches. There are plenty of features here to improve the visuals too, such as Apple's True Tone display tech, which subtly tweaks the display depending on what environment you're in.

The iPad Air 4 is the first Apple tablet with a fingerprint sensor built into a button on the side, which is an intriguing move from Apple. Other slates from the company have had fingerprint sensors built into the home button on the front, but we haven't seen side-mounted scanners from Apple before.

There's a USB-C port here, as on the iPad Pro models, which should make charging and data sending much faster, and it'll be welcomed by professionals who want to plug external monitors or hard drives into the tablet

The iPad Air 4 works with the Magic Keyboard and the second-generation Apple Pencil – that makes it the first tablet outside the Apple Pro range to use this newer stylus. It attaches to the top of the tablet magnetically for storage and charging.

iPad Air specs and features

The chipset inside the iPad Air 4 is the A14 Bionic, which we're expecting to see in the iPhone 12. It's the first 5nm chipset in an Apple mobile tablet, and it sounds like it'll make the new Air the fastest tablet Apple has put out yet – we'll run some benchmark tests as soon as we get the slate into our labs, to see if that's the case.

Apple says the graphics performance of the tablet is "30% faster", although it hasn't said what that's compared to. It does claim that the iPad Air 4 is twice as fast as your average HP laptop though.

Apple rarely announces the battery capacity of its devices when they're launched – we'd expect to find that out when the iPad Air is available to buy, as people will inevitably take it apart to see what's inside.

Finally there's a 7MP f/2.2 front-facing camera and 12MP f/1.8 rear camera, which should be good enough for video calls and other forms of communication.

Chapter 1
Set up and get started

Turn on and set up iPad

Turn on and set up your new iPad over an Internet connection. You can also set up an iPad by connecting it to your computer. If you have another iPhone, iPad, iPod touch, or an Android device, you can transfer your data to your new iPad.

Prepare for setup

To make setup as smooth as possible, have the following items available:

- An Internet connection through a Wi-Fi network (you may need the name and password of the network) or cellular data service through a carrier (Wi-Fi + Cellular models)

- Your Apple ID and password; if you don't have an Apple ID, you can create one during setup

- Your credit or debit card account information, if you want to add a card to Apple Pay during setup

- Your previous iPad or a backup of your device, if you're transferring your data to your new device

- Your Android device, if you're transferring your Android content

Turn on and set up your iPad

- Press and hold the top button until the Apple logo appears.

- If iPad doesn't turn on, you might need to charge the battery.

Do one of the following:

- Tap Set Up Manually, then follow the onscreen setup instructions.

- If you have another iPhone, iPad, or iPod touch with iOS 11, iPadOS 13, or later, you can use Quick Start to automatically set up your new device. Bring the two devices close together, then follow the onscreen instructions to securely copy many of your settings, preferences, and iCloud Keychain. You can then restore the rest of your data and content to your new device from your iCloud backup.

- Or, if both devices have iOS 12.4, iPadOS 13, or later, you can transfer all your data wirelessly from your previous device to your new one. Keep your devices near each other and plugged into power until the migration process is complete.

- You can also transfer your data using a wired connection between your devices.

- If you're blind or have low vision, triple-click the Home button to turn on VoiceOver, the screen reader.

Set up cellular service on iPad (Wi-Fi + Cellular models)

If you have a Wi-Fi + Cellular model, you can sign up for a cellular data plan. This helps you stay connected to the Internet when you're away from a Wi-Fi network. You can set up a cellular data plan with any of the following:

- eSIM

- Embedded Apple SIM or Apple SIM card

- Third-party nano-SIM (from a cellular provider)

Set up your cellular plan with eSIM

On models that support eSIM, you can activate the cellular service from your iPad. You may also be able to travel abroad with iPad and sign up for cellular service with a local carrier in the region you're visiting. This isn't available in all regions, and not all carriers are supported. Go to Settings > Cellular Data.

Do one of the following:

- To set up the first cellular plan on your iPad, select a carrier, then follow the onscreen instructions.

- To add another cellular plan to your iPad, tap Add a New Plan.

- To scan a QR code provided by your carrier, tap Other. Position iPad so that the QR code provided by your carrier appears in the frame, or enter the details manually. You may be asked to enter a confirmation code provided by your carrier.

Alternatively, you can activate your cellular plan through your carrier's app (if supported). Go to the App Store, download your carrier's app, then use the app to purchase a cellular plan.

You can store more than one eSIM on your iPad, but you can use only one eSIM at a time. To switch eSIMs, go to Settings > Cellular Data, then tap the plan you want to use (below Cellular Plans).

Set up your cellular plan with an embedded Apple SIM or Apple SIM card

On models with an embedded Apple SIM or Apple SIM card, you can activate the cellular service from your iPad. You may also be able to travel abroad with iPad and sign up for cellular service with a local carrier in the region you're visiting. This isn't available in all regions, and not all carriers are supported. Go to Settings > Cellular Data.

- Tap Add a New Plan, then follow the onscreen instructions. You can choose a carrier and a plan, or you can add your iPad to an existing plan.

Install a nano-SIM

You can install an Apple SIM card or a nano-SIM provided by a carrier.

- Insert a paper clip or SIM eject tool (not included) into the small hole of the SIM tray, then push in toward iPad to eject the tray.

- Remove the tray from iPad.

- Place the nano-SIM in the tray. The angled corner determines the correct orientation.

- Insert the tray back into iPad.

- If you previously set up a PIN on the nano-SIM, carefully enter the PIN when prompted.

WARNING: Never try to guess a SIM PIN. An incorrect guess can permanently lock your SIM, and you won't be able to use cellular data through your carrier until you get a new SIM.

Cellular data requires a wireless data plan. If you're using a third-party nano-SIM, contact your carrier to set up service.

Manage your cellular data service

Go to Settings > Cellular Data. Do any of the following:

- Restrict all data to Wi-Fi: Turn off Cellular Data.

- Turn on or off LTE and roaming: Tap Cellular Data Options.

- Turn on Personal Hotspot: Tap Set Up Personal Hotspot (available from certain carriers), then follow the onscreen instructions.

- Manage your cellular account: Tap Manage [account name] or Carrier Services.

Manage Apple ID and iCloud settings on iPad

Your Apple ID is the account you use to access Apple services such as the App Store, the iTunes Store, Apple Books, Apple Music, FaceTime, iCloud, iMessage, and more.

Use iCloud to securely store your photos, videos, documents, music, apps, and more—and keep them updated across all your devices. With iCloud, you can easily share photos, calendars, locations, and more with friends and family. You can even use iCloud to help you find your iPad if you lose it.

iCloud provides you with a free email account and 5 GB of storage for your mail, documents, photos and videos, and backups. Your purchased music, apps, TV shows, and books don't count against your available storage space. You can upgrade your iCloud storage right from iPad.

Note: Some iCloud features have minimum system requirements. The availability of iCloud and its features varies by country or region.

Sign in with your Apple ID

If you didn't sign in during setup, do the following:

- Go to Settings .
- Tap Sign in to your iPad.
- Enter your Apple ID and password.
- If you don't have an Apple ID, you can create one.
- If you protect your account with two-factor authentication, enter the six-digit verification code.

Change your Apple ID settings

Go to Settings > [your name]. Do any of the following:

- Update your contact information
- Change your password
- Manage Family Sharing

Change your iCloud settings

- Go to Settings > [your name] > iCloud.

Do any of the following:

- See your iCloud storage status.
- Upgrade your iCloud storage—tap Manage Storage > Change Storage Plan.

- Turn on the features you want to use, such as Photos, Mail, Contacts, and Messages.

Move content manually from your Android device to your iOS device

Here are some tips for transferring your contacts, photos, music, documents, and more from your Android device to your new iPhone, iPad, or iPod touch.

You can also use the Move to iOS app to automatically transfer your Android content to your new iOS device. If you can't use the app, you can move your content manually instead.

Mail, contacts, and calendars

iOS works with email providers like Google, Microsoft Exchange, Yahoo, and more, so you can probably keep the email, contacts, and calendars that you have now. To get started, add each of your email accounts to your iPhone. Then go to Settings > Passwords & Accounts.

Photos and videos

To move photos and videos from your Android device to your iOS device, use a computer with iTunes:

- Connect your Android to your computer and find your photos and videos. On most devices, you can find these files in DCIM > Camera. On a Mac, install

Android File Transfer, open it, then go to DCIM > Camera.

- Choose the photos and videos that you want to move and drag them to a folder on your computer.

- Disconnect your Android and connect your iPhone to your computer.

- Open iTunes on your computer and sync your Photos to your iPhone. You can find your photos and videos on your iPhone in Photos > Albums.

You can also use iCloud Photos to keep your photos and videos in iCloud, so you can access your library from any device, anytime you want.

Music

When you switch to iPhone, you can bring your music with you. Just use a computer with iTunes to transfer the music. If you use a streaming music app, go to the App Store, get the app, then sign in with your user name and password. If you use Apple Music, just sign in on your iOS device.

To move music from your Android device to your iOS device, use a computer with iTunes:

- Connect your Android device to your computer and find your music. On most devices, you can find these files in Music. On a Mac, install Android File Transfer, open it, then go to Music.

- Select the songs that you want to move and drag them to a folder on your computer.

- Disconnect your Android device and connect your iPhone to your computer.

- Open iTunes on your computer, go to your Library, and click Music.

- Open the folder where you put your songs and drag them to the Music view in iTunes.

- Select your iOS device and click Music. You can choose to sync your entire library or select only the songs or artists that you just added

- Click Sync. You can find your music on your iPhone in the Music app.

Books and PDFs

To move eBooks from your Android device, you can either import them to your iOS device or access them through apps like Kindle, Nook, Google Play Books, and others. To access books from an app, go to the App Store, get the app, then sign in with your username and password.

To move ePub books and PDFs from your Android to your iOS device, use a computer with iTunes:

- Connect your Android device to your computer and find your books and PDFs. On most devices, you can find these files in Documents. On a Mac, install Android File Transfer, open it, then go to Documents.

- Select the books and PDFs that you want to move and drag them to a folder on your computer.

- Disconnect your Android device and connect your iPhone to your computer.

- Drag the books and PDFs into your library. On a Mac, go to Books > List, and drag your books there. On a PC, go to iTunes > Books.

- Open iTunes on your computer and sync your ePub books and PDFs. You can find your ePub books and PDFs on your iPhone inBooks > All Books.

Documents

If you store documents in the cloud or another service like Dropbox, Google Drive, or Microsoft OneDrive, you can download the app from the App Store, then sign in. You can also bring all your files together with the Files app.

Whether your files are on your iOS device, in iCloud Drive, or on another service like Dropbox or Box, you can easily browse, search, and organize your files all in one place.

The iOS apps for Pages, Numbers, and Keynote work with several file types, including Microsoft Office documents. If you don't use the cloud to transfer your documents, get the apps:

- Go to the App Store on your iPhone and install Pages, Numbers, and Keynote.

- Connect your Android to your computer and find your documents. On most devices, you can find these files in Documents. On a Mac, install Android File Transfer, open it, then go to Documents.

- Select the documents that you want to move and drag them to a folder on your computer.

- Open iTunes on your computer and sync your documents to your iOS device.

Apps

Most apps that you use on your Android device are available in the App Store. Go to the App Store, search for the apps that you have now, and install them.

How to download iPadOS on an Apple tablet

iPadOS represents a bit of a shift for Apple, as it continues to differentiate how the operating system on the iPad works, feels, and functions. Now, iPadOS is available to the public, bringing with it a number of great features, including mouse support, a revamped home screen, and more.

There are plenty of features to check out, but to see them for yourself you'll need to update your iPad. Here's how to download iPadOS.

Compatible devices

Before installing iPadOS, you'll need to make sure your iPad is compatible with the new operating system. Apple is known for supporting devices for a long time, but some older models won't get the update, so it's worth checking the list below to see if your iPad will get support.

- iPad Air 2 and 3

- iPad Mini 4 and 5

- iPad (6th and 7th generations)

- 9.7-inch iPad Pro

- 10.5-inch iPad Pro

- 11-inch iPad Pro

- 12.9-inch iPad Pro

Installing iPadOS onto your device is really super simple. Here's how to get iPadOS on your iPad. Note, if you think you might not like iPadOS and might want to roll back to iOS 12, then it's worth creating a backup before you start. Check out the instructions below on creating a backup.

- Open the Settings app.

- Head to General > Software Update.

- Your iPad will check for updates and you should get a notification telling you that iPadOS is ready to install. Tap Download and Install.

- It may take a few minutes to download and install the update, and you won't be able to use your iPad during the update process.

Back up your iPad

Think you might want to roll back to iOS 12 after updating your device? In that case, you should make a backup before upgrading. There are two ways to backup your iPad — using iCloud, or through iTunes.

Backing up using iCloud

Backing up your iPad using iCloud is the easiest method. Here's how to do it for yourself.

- Make sure you're connected to a Wi-Fi network.

- Open the Settings app, press your name, then tap iCloud.

- Scroll down to iCloud Backup, then tap Back Up Now.

If you're unsure as to whether the backup is complete, you can head to Settings, then tap iCloud > iCloud Storage > Manage Storage, then tap on the device on the list.

Backing up on a Mac running MacOS Catalina

MacOS Catalina no longer has iTunes, so backing up your device on a Mac is a little different than it used to be.

If you're running MacOS Catalina, you'll instead use the Finder app. Here's how it's done.

- Connect your iPad to your Mac.

- Follow the onscreen instructions — you may need to enter a PIN code or tap Trust This Computer.

- Open the Finder app and select your iPad in the sidebar.

- Press the General tab, then tap Back Up Now to manually back up your iPad.

Backing up on a Mac or PC with iTunes

If you have a Mac with Mojave or older, or a PC with iTunes, then you'll use iTunes to back up your iPad. Here's how to do it.

- Make sure you have the latest version of iTunes, then connect your iPad.

- Follow the onscreen instructions — you may need to enter a PIN code or tap Trust This Computer.

- Open iTunes and select your iPad.

- Press the Back Up Now button to save your data.

Restore from an iCloud backup

Here's how to restore your device from an iCloud backup from before you updated to iOS 12.

- On the Apps & Data screen, tap Restore from iCloud Backup and sign in to iCloud.

- Tap Choose Backup and choose the backup you made before installing iPadOS.

- Make sure you select the right backup — if you've had your device for more than a day, you may have another backup from when you were already on iPadOS.

Restore from an iTunes backup

Did you make an iTunes backup instead? Here's how to restore from an iTunes backup.

- Tap Restore from iTunes Backup on the Apps & Data screen.

- Open iTunes on your computer, make sure your device is connected through a cable, then tap Trust This Computer.

- Select your device in iTunes, then press Summary and hit the Restore Backup button.

- Pick the backup from when your device was still running iOS 12.

- Keep your iPad connected to your computer until after it finishes syncing.

Chapter 2
Basics Guide

Wake and unlock iPad

iPad turns off the display to save power, locks for security, and goes to sleep when you're not using it. You can quickly wake and unlock iPad when you want to use it again. To wake iPad, do one of the following:

- Press the top button.

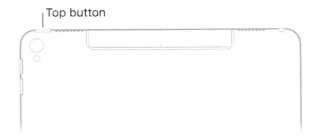

- Tap the screen. Or, on **supported models**, you can tap the screen with Apple Pencil to wake iPad and open Notes.

Unlock iPad with Face ID

1. On **supported models**, tap the screen, then glance at your iPad. The lock icon animates from closed to open to indicate that iPad is unlocked.

2. Swipe up from the bottom of the screen to view the Home screen.

To lock iPad again, press the top button. iPad locks automatically if you don't touch the screen for a minute or so. However, if Attention Aware Features is turned on in Settings 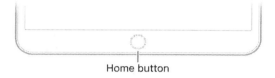 > Face ID & Passcode, iPad won't dim or lock as long as it detects attention.

Unlock iPad with Touch ID

On **supported models**, press the Home button using the finger you registered with Touch ID.

Home button

To lock iPad again, press the top button. iPad locks automatically if you don't touch the screen for a minute or so.

- On iPad Air (4th generation), press the top button (Touch ID) using the finger you registered with Touch ID.

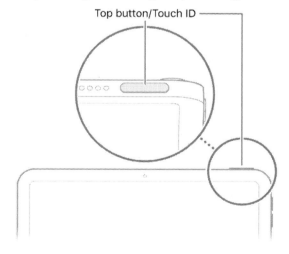

Top button/Touch ID

To lock iPad again, press the top button. iPad locks automatically if you don't touch the screen for a minute or so.

Unlock iPad with a passcode

1. Swipe up from the bottom of the Lock screen or press the Home button.

2. Enter the passcode (if you set up iPad to require a passcode).

To lock iPad again, press the top button. iPad locks automatically if you don't touch the screen for a minute or so.

View previews and quick actions menus on iPad

On the **Home** screen, in **Control Center**, and in apps, you can see previews, open quick actions menus, and more.

- In **Photos**, touch and hold an image to preview it and see a list of options.

- In **Mail**, touch and hold a message in a mailbox to preview the message contents and see a list of options.

- On the Home screen, touch and hold an app icon briefly to open a quick actions menu. If the icons start to jiggle, tap **Done** at the top right or press the **Home button** (models with a Home button), then try again.

- Open Control Center, then touch and hold an item like Camera or the brightness control to see options.

- On the Lock screen, touch and hold a notification briefly to respond to it.

- When typing, touch and hold the Space bar with one finger to turn your keyboard into a trackpad.

Explore the iPad Home screen and open apps

Get to know the Home screen and apps on your iPad. The Home screen shows all your apps organized into pages. More pages are added when you need space for apps.

1. To go to the Home screen, swipe up from the bottom edge of the screen or press the Home button.

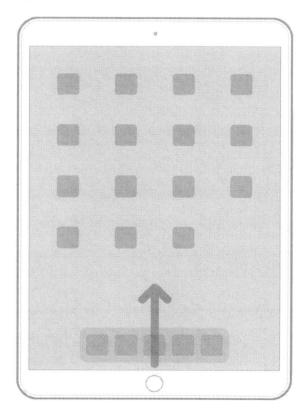

2. Swipe left or right to browse apps on other Home screen
   pages.

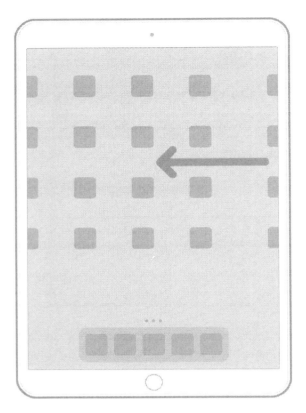

3. To open an app, tap its icon on the Home screen.

4. To return to the first Home screen page, swipe up from the
   bottom edge of the screen or press the Home button.

Change common iPad settings

Use Settings 🔘 (located on the Home screen) to configure and
customize your iPad settings. You can set your language and
region, change the name of your iPad, choose different sounds
for notifications, and much more.

The settings for specific apps are explained in the chapters for those apps. The following sections give some examples of common settings, including how to find them.

Tap Settings to change iPad settings (volume, display brightness, and more).

Find settings

Go to Settings , swipe down from the left side of the screento reveal the search field, enter a term—"iCloud," for example— then tap a setting on the left side of the screen.

Use iPad to search

Search on iPad is the best place to start all your searches. Search can help you find apps and contacts, search inside of apps like Mail and Messages, find and open webpages, and quickly start a web search.

You can choose which apps you want to be included in search results. Search offers suggestions and updates results as you type.

Choose which apps to include in Search

1. Go to Settings  > Siri & Search.

2. Scroll down, tap an app, then turn Show in Search on or off.

Search with iPad

1. Swipe down from the middle of the Home Screen.

2. Tap the search field, then enter what you're looking for.

3. Do any of the following:

   - *Hide the keyboard and see more results on the screen:* Tap Go.

   - *Open a suggested app:* Tap it.

   - *Get more information about a search suggestion:* Tap it, then tap one of the results to open it.

   - *Start a new search:* Tap ⊗ in the search field.

Turn off Suggestions in Search

Go to Settings 🔘 > Siri & Search, then turn off Suggestions in Search.

Turn off Location Services for suggestions

1. Go to Settings 🔘 > Privacy > Location Services.

2. Tap System Services, then turn off Location-Based Suggestions.

Search in apps

Many apps include a search field or a search button so you can find something within the app. For example, in the Maps app, you can search for a specific location.

1. In an app, tap the search field or button $\mathcal{Q}$ (if there is one).

   If you don't see a search field or button, swipe down from the top.

2. Type your search, then tap Search.

Add a dictionary

On iPad, you can add dictionaries, which can be used in searches.

1. Go to Settings 🔘 > General > Dictionary.

2. Select a dictionary.

Set the date and time

By default, the date and time, visible on the Lock screen, are set automatically based on your location. If they're incorrect, you can adjust them.

1. Go to Settings ⚙ > General > Date & Time.

2. Turn on either of the following:

- **Set Automatically**: iPad gets the correct time over the network and updates it for the time zone you're in. Some networks don't support network time, so in some regions iPad may not be able to automatically determine the local time.

- **24-Hour Time**: (not available in all regions) iPad displays the hours from 0 to 23.

To change the default date and time, turn off Set Automatically, then change the date and time displayed.

Set the language and region

1. Go to Settings ⚙ > General > Language & Region.

2. Set the following:

- The language for iPad

- The region

- The calendar format

- The temperature unit (Celsius or Fahrenheit)

To add a keyboard for another language, go to Settings > General > Keyboard > Keyboards, then tap Add New Keyboard.

Change the name of your iPad

The name of your iPad is used by iCloud, AirDrop, your Personal Hotspot, and your computer.

1. Go to Settings ⚙ > General > About > Name.

2. Tap ⊗, enter a new name, then tap Done.

Set up mail, contacts, and calendar accounts

In addition to the apps that come with iPad and that you use with **iCloud**, iPad works with Microsoft Exchange and many of the most popular Internet-based mail, contacts, and calendar services.

1. Go to Settings ⚙ > Passwords & Accounts > Add Account.

2. To add a mail account, tap an email service—for example, Google, Yahoo, or Aol.com—then enter your email account information.

3. To add a contacts or calendar account, tap Other, then do any of the following:

   - Contacts using an LDAP or CardDAV account, if your company or organization supports it.

- Calendars using a CalDAV calendar account; you can also subscribe to iCalendar (.ics) calendars or import them from Mail.

- *Add a contacts account:* Tap Add LDAP Account or Add CardDAV Account (if your company or organization supports it), then enter your information.

- *Add a calendar account:* Tap Add CalDAV Account, then enter your information.

- *Subscribe to iCal (.ics) calendars:* Tap Add Subscribed Calendar, then enter the URL of the .ics file to subscribe to; or import an .ics file from Mail.

Change or lock the screen orientation on iPad.

Many apps give you a different view when you rotate iPad.

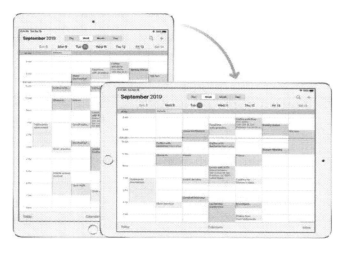

Lock or unlock the screen orientation

- You can lock the screen orientation so that it doesn't change when you rotate iPad.

32

- Open Control Center, then tap 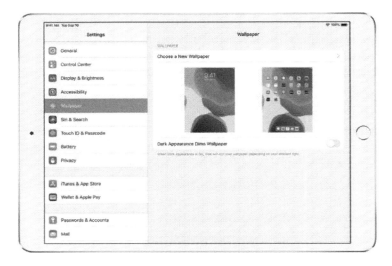.

- When the screen orientation is locked, appears in the status bar.

Change the wallpaper on iPad

On iPad, choose an image or photo as wallpaper for the Lock screen or Home screen. You can choose from dynamic and still images.

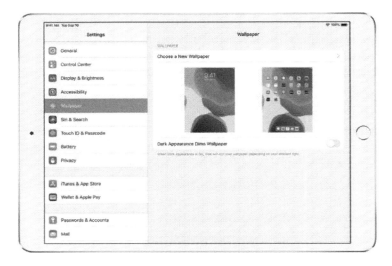

1. Go to Settings > Wallpaper > Choose a New Wallpaper.

2. Do one of the following:

  - Choose a preset image from a group at the top of the screen (Dynamic, Stills, and so on).

    Wallpaper marked with ◑ changes appearance when Dark Mode is turned on.

- Select one of your own photos (tap an album, then tap the photo).

  To reposition your selected image, pinch open to zoom in on it, then drag the image to move it. Pinch closed to zoom back out.

  Tap Set, then choose one of the following:

- Set Lock Screen

- Set Home Screen

- Set Both

You may be able to make your wallpaper move when you change the viewing angle of your screen by turning on Perspective Zoom when you choose new wallpaper. To turn on the Perspective Zoom option for wallpaper you've already set, go to Settings > Wallpaper, tap the image of the Lock screen or Home screen, then tap Perspective.

Note: The Perspective Zoom option doesn't appear for all wallpaper choices, and it doesn't appear if Reduce Motion (in Accessibility settings) is turned on.

Adjust the iPad screen brightness and color

On iPad, dim the screen to extend battery life, set Dark Mode, and use Night Shift. On supported models, use True Tone to automatically adapt the color and intensity of the display to match the light in your environment.

Turn Dark Mode on or off

Dark Mode gives the entire iPad experience a dark color scheme that's perfect for low-light environments. You can turn on Dark Mode from Control Center or set it to turn on automatically at night (or on a custom schedule) in Settings. With Dark Mode turned on, you can use your iPad while, for example, reading in bed, without disturbing the person next to you.

Do any of the following:

- Open Control Center, touch and hold ☼, then tap ◑ to turn Dark Mode on or off.

- Go to Settings ⚙ > Display & Brightness, then select Dark to turn on Dark Mode or select Light to turn it off.

Schedule Dark Mode to turn on and off automatically

1. Go to Settings ⚙ > Display & Brightness.

2. Turn on Automatic, then tap Options.

3. Select either Sunset to Sunrise or Custom Schedule.

   If you choose Custom Schedule, tap the options to schedule the times you want Dark Mode to turn on and off.

   If you select Sunset to Sunrise, iPad uses the data from your clock and geo-location to determine when it's nighttime for you.

Adjust the screen brightness manually

To make your iPad screen dimmer or brighter, do one of the following:

- Open Control Center, then drag ☼.

- Go to Settings ⚙ > Display & Brightness, then drag the slider.

Adjust the screen brightness automatically

iPad adjusts the screen brightness for current light conditions using the built-in ambient light sensor.

1. Go to Settings ⚙ > Accessibility.

2. Tap Display & Text Size, then turn on Auto-Brightness.

Turn True Tone on or off

On supported models, turn on True Tone to automatically adapt the color and intensity of the display to match the light in your environment. Do any of the following:

36

- Open Control Center, touch and hold ☼, then tap ☀ to turn True Tone on or off.

- Go to Settings ◎ > Display & Brightness, then turn True Tone on or off.

Turn Night Shift on or off

You can turn on Night Shift manually, which is helpful when you're in a darkened room during the day.

- Open Control Center, touch and hold ☼, then tap ☾.

Schedule Night Shift to turn on and off automatically

Use Night Shift to shift the colors in your display to the warmer end of the spectrum at night and make viewing the screen easier on your eyes.

1. Go to Settings ◎ > **Display & Brightness** > **Night Shift.**

2. Turn on **Scheduled**.

3. To adjust the color balance for **Night Shift**, drag the slider below Color Temperature toward the warmer or cooler end of the spectrum.

4. Tap **From**, then select either **Sunset** to **Sunrise** or **Custom Schedule**.

   If you choose Custom Schedule, tap the options to schedule the times you want Night Shift to turn on and off.

If you select Sunset to Sunrise, iPad uses the data from your clock and geo-location to determine when it's nighttime for you.

Note: The Sunset to Sunrise option isn't available if you turned off Location Services in Settings ⚙ > Privacy, or if you turned off Setting Time Zone in Settings ⚙ > Privacy > Location Services > System Services.

Magnify the iPad screen with Display Zoom

On iPad Pro (12.9-inch), you can magnify the screen display.

1. Go to Settings ⚙ > Display & Brightness.

2. Tap View (below Display Zoom), choose Zoomed, then tap Set.

Chapter 3
How to use Apps

Switch between apps on iPad

Use the Dock, the App Switcher, or a gesture to quickly switch from one app to another on your iPad. When you switch back, you can pick up right where you left off.

Use app clips on iPad

An app clip is a small part of an app that lets you do a task quickly, without downloading and installing the full app. You can discover app clips in Safari, Maps, Messages, or in the real world when you rent a bike, pay for parking, or order food.

Get an app clip

When you discover an app clip, open it in one of the following ways:

- Tap the app clip link in Safari, Maps, or Messages.

- Using the iPad camera, scan the QR code shown at the physical location, such as a restaurant or payment terminal.

The app clip card appears at the bottom of the screen.

Remove app clips

Go to Settings > App Clips, then tap Remove All App Clips.

Subscribe to Apple Arcade on iPad

In the App Store app , you can subscribe to Apple Arcade to enjoy unlimited access to a curated collection of games on iPhone, iPad, iPod touch, Mac, and Apple TV.

If you use Family Sharing, up to five other family members can share the subscription for no additional charge.

Note: Apple Arcade isn't available in all countries or regions.

Get Apple Arcade

1. In the App Store, tap Arcade, then do one of the following:

- *Start a free one-month subscription (if eligible):* Tap Try It Free.

- *Start a monthly subscription:* Tap Subscribe.

  Review the subscription details, then confirm with Face ID, Touch ID, or your Apple ID.

Cancel your Apple Arcade subscription

1. In the App Store, tap 🄰 or your profile picture at the top right, then tap Subscriptions.

2. Tap Apple Arcade, then tap Cancel Subscription.

After you cancel your subscription, you won't be able to play any Apple Arcade games, even if you downloaded them to your device. Delete the apps if you don't need them anymore.

You can resubscribe to play Apple Arcade games again and regain access to your gameplay data. If you wait too long, some of your gameplay data might not be supported after you resubscribe.

Play games on iPad

In the App Store app 🄰, you can discover new games and play with your friends using Game Center.

Find and download games

1. Tap any of the following tabs:

- *Games:* Explore new releases, see the top charts, or browse by category.

- *Arcade:* See the games available in Apple Arcade.

- *Search:* Enter what you're looking for, then tap Search on the keyboard.

To buy a game, tap the price or, if the game is free, tap Get. If the game is included with your Apple Arcade subscription, tap Play.

If you see ☁ instead of a price, you already purchased the game, and you can download it again without a charge.

If required, authenticate your Apple ID with Face ID, Touch ID, or your passcode to complete your purchase.

Play Apple Arcade games on your other Apple devices

All of the games in Apple Arcade on iPad are also available in Apple Arcade on other devices. If you subscribe to Apple Arcade, you can access your game progress on devices where you're signed in with your Apple ID.

Play with your friends on Game Center

You can send friend requests, manage your public profile, and track your high scores across your Apple devices using Game Center.

1. Go to Settings ⚙ > Game Center, then sign in with your Apple ID.

2. To create a Game Center profile, do any of the following:

   • *Choose a nickname:* Tap Nickname, then enter a name or choose one of the suggestions. Your friends see your nickname when you play games together.

   • *Personaliz your profile picture:* Tap Edit at the top, then create a new Memoji, use an existing Memoji, or customize how your initials appear.

   To add friends, tap Friends, tap Add Friends, then enter their phone number or Apple ID, or tap ⊕ to invite someone in your Contacts list.

   To accept a friend request, the recipient must click the link in the text message on their iPhone, iPad, iPod touch, or Mac that meets the minimum system requirements for Apple Arcade.

In your list of friends, tap a friend to see games they recently played and their achievements. You can also report a user for cheating, an inappropriate picture or nickname, or another problem. To remove a friend, tap Remove Friend.

Set Game Center restrictions

You can set restrictions for multiplayer games, adding friends, private messaging, and more.

1. Go to Settings > Screen Time > Content & Privacy Restrictions, then turn on Content & Privacy Restrictions.

2. Tap Content Restrictions, scroll down to Game Center, then set restrictions.

Install and manage fonts on iPad

You can download fonts from the App Store app , then use them in documents you create on iPad.

1. After you download an app containing fonts from the App Store, open the app to install the fonts.

2. To manage installed fonts, go to Settings > General, then tap Fonts.

Find and download apps purchased by you or family members

1. Tap or your profile picture at the top right, then tap Purchased.

2. If you set up Family Sharing, tap My Purchases or choose a family member to view their purchases.

   Note: You can see purchases made by family members only if they choose to share their purchases. Purchases made with Family Sharing may not be accessible after the family member leaves the family group.

3. Find the app you want to download (if it's still available in

the App Store), then tap ⛅︎⬇︎ .

Manage your subscriptions

Tap 👤 or your profile picture at the top right, then tap Subscriptions.

Change your App Store settings

Go to Settings ⚙︎ > App Store, then do any of the following:

- *Automatically download apps purchased on your other Apple devices:* Below Automatic Downloads, turn on Apps.

- *Automatically update apps:* Turn on App Updates.

- *Control the use of cellular data for app downloads:* (Wi-Fi + Cellular models) To allow downloads to use cellular data, turn on Automatic Downloads (below Cellular Data). To choose whether you want to be asked for permission for downloads over 200 MB or all apps, tap App Downloads.

- *Automatically play app preview videos:* Turn on Video Autoplay.

- *Automatically remove unused apps:* Turn on Offload Unused Apps. You can reinstall an app at any time if it's still available in the App Store.

Set content restrictions and prevent in-app purchases

After you turn on content and privacy restrictions, do the following.

1. Go to Settings ⚙ > Screen Time > Content & Privacy Restrictions > Content Restrictions.

2. Set restrictions such as the following:

   - *iTunes & App Store Purchases:* Control app installations, app deletions, and in-app purchases.

   - *Apps:* Restrict apps by age ratings.

   - *App Clips:* Prevent app clips from opening.

Open an app from the dock

From any app, swipe up from the bottom edge of the screen and pause to reveal the Dock, then tap the app you want to use. Favorite apps are on the left side of the Dock, and suggested apps—like the ones you opened recently and ones open on your iPhone or Mac—appear on the right side of the Dock.

Favorite apps   Suggested apps

Use the App Switcher

1. To see all your open apps in the App Switcher, do one of the following:

- Swipe up from the bottom edge and pause in the center of the screen.

- Double-click the Home button (models with the Home button).

2. To browse the open apps, swipe right, then tap the app or Split View workspace you want to use.

Switch between open apps

- Swipe left or right with four or five fingers.

- Swipe left or right with one finger along the bottom edge of the screen. (On models with a Home button, perform this gesture with a slight arc.)

To turn off the multi finger swipe gesture, go to Settings
> General > Multitasking & Dock.

Move and organize apps on iPad

Rearrange the apps on the Home screen, organize them in folders, and move them to other pages (or screens). You can also reorder your pages.

Move apps around the Home screen, into the Dock, or to other pages

1. Touch and hold an app on the Home screen until the app icons jiggle.

2. Drag an app to one of the following locations:

- Another location on the same page

- The Dock at the bottom of the screen

- Another page—drag the app to the right edge of the screen. You might need to wait a second for the new page to appear. The dots above the Dock shows how many pages you have, and which one you're viewing.

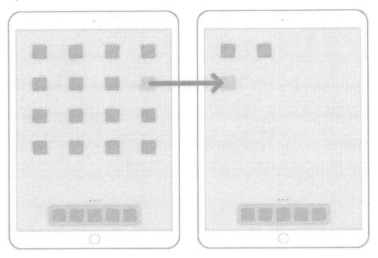

3. When you're done, swipe up   from the bottom edge of the screen or press the Home button (models with the Home button).

Create folders and organize your apps

You can group your apps in folders to help you find them more easily on the Home screen.

1. Touch and hold any app on the screen until the app icons jiggle.

2. To create a folder, drag an app onto another app.

3. Drag other apps into the folder.

4. You can have multiple pages of apps in the folder.

5. To rename the folder, tap the name field, then enter the new name.

6. When you're done, swipe up from the bottom edge of the screen or press the Home button (models with the Home button).

To delete a folder, drag all the apps out of the folder. The folder is automatically deleted.

Reset the Home screen and apps to their original layout

1. Go to Settings ⚙ > General > Reset.

2. Tap Reset Home Screen Layout. Any folders you've created are removed, and apps you've downloaded are alphabetically ordered after apps that came with your iPad.

Remove apps from iPad

You can easily remove apps from your iPad. If you change your mind, you can download the apps again later.

Remove apps from the Home screen

1. Touch and hold any app on the screen until the app icons jiggle.

2. Tap the Close button on the app you want to remove, then tap Delete.

3. When you're done, swipe up from the bottom edge or press the Home button (models with the Home button).

In addition to removing third-party apps, you can remove the following built-in Apple apps that came with your iPad:

- Books

- Calendar

- Contacts (Contact information remains available through Messages, Mail, FaceTime, and other apps. To remove a contact, you must restore Contacts.)

- FaceTime

- Files

- Find My

- Home

- iTunes Store

- Mail

- Maps

- Measure

- Music

- News

- Notes

- Photo Booth

- Podcasts

- Reminders

- Shortcuts

- Stocks

- Tips

- TV

- Voice Memos

Note: When you remove a built-in app from your Home screen, you also remove any related user data and configuration files. Removing built-in apps from your Home screen can affect other system functionality.

Keep your favorite apps readily available on iPad

On iPad, you can keep your favorite apps handy in Control Center or Today View. In Control Center, shortcuts give you quick access to apps like Notes or Voice Memos. In Today View, widgets provide timely information from your favorite apps at a glance.

You can also perform common app functions from the Home screen. On the Home screen, touch and hold an app icon to open a quick actions menu.

Customize Control Center to include your favorite apps

You can add shortcuts to many apps such as Notes, Voice Memos, and more.

1. Go to Settings > Control Center > Customize Controls.

2. Tap the Insert button ⊕ next to each app you want to add.

Add widgets in Today View

1. Get information from your favorite apps at a glance. Choose from Maps Nearby, Calendar, Notes, News, Reminders, and more.

2. From the Home screen, swipe right to open Today View.

3. Scroll to the bottom, then tap Edit.

4. Tap the Insert button ⊕ next to each app you want to add, then tap **Done**.

Add widgets to the iPad Home Screen

Today View widgets show you current information from your favorite apps at a glance—today's headlines, weather, calendar events, and more. You can add these widgets to your Home Screen to keep this information at your fingertips.

You can keep Today View widgets
on your Home Screen.

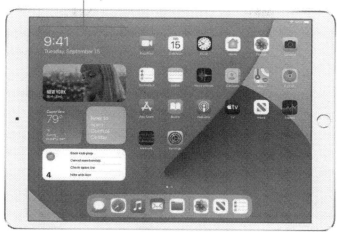

Open Today View

Swipe right from the left edge of the Home Screen or the Lock Screen.

Keep Today View widgets on your Home Screen

You can keep Today View widgets on your Home Screen next to your apps when iPad is in landscape orientation.

1. With iPad in landscape orientation, open Today View, then touch and hold the Home Screen background until the apps jiggle.

2. Turn on Keep On Home Screen, then tap Done.

Tip: A widget called a Smart Stack (one that has dots next to it) is a set of several widgets that uses information such as the time, your location, and activity to display the most relevant widget at the appropriate time in your day. You can swipe through a Smart Stack to see the widgets it contains.

Add widgets from the widget gallery

1. Open Today View, then touch and hold the Home Screen
   background until the apps begin to jiggle.

2. Tap 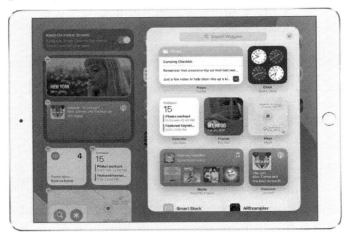 at the top of the screen to open the widget
   gallery.

3. Scroll or search to find the widget you want, tap it, then
   swipe through the size options.

   The different sizes display different information.

4. When you see the size you want, tap Add Widget, then tap
   Done.

Tip: A widget called a Smart Stack (one that has dots next to
it) is a set of several widgets that uses information such as the
time, your location, and activity to display the most relevant
widget at the appropriate time in your day. You can swipe
through a Smart Stack to see the widgets it contains.

Remove or rearrange widgets

1. In landscape orientation, **open Today View** on the Home Screen.

2. Touch and hold the Home Screen background until the apps begin to jiggle, then do any of the following:

   - *Remove a widget:* Tap on the widget, then tap Remove.

   - *Rearrange the widgets:* Drag a widget to a new location in Today View.

Edit a widget

You can customize most widgets so they display the information you want. For example, you can edit a Weather widget to display the forecast for your location or a different area. Or you can customize a Smart Stack to rotate automatically through its widgets based on things like your activity, the time of day, and so on.

1. On your Home Screen, touch and hold a widget to open a quick actions menu.

2. Tap Edit Widget if it appears (or Edit Stack, if it's a Smart Stack), then choose options.

   For example, for a Weather widget, you can tap Location, then select a location for your forecast.

For a Smart Stack, you can turn Smart Rotate off or on and reorder the widgets in the stack by dragging ☰ next to them.

3. Tap the Home Screen background.

Perform quick actions from the Home screen

On the Home screen, touch and hold app icons to open quick actions menus.

For example:

- Touch and hold Camera 📷, then choose Take Selfie.

- Touch and hold Maps 🗺, then choose Send My Location.

- Touch and hold Notes 📝, then choose New Note.

Note: If you touch and hold an app icon for too long before choosing a quick action, the icons begin to jiggle. Tap **Done** or press the Home button (models with a Home button), then try again.

Draw in apps with Markup on iPad

In supported apps such as Messages, Mail, Notes, and Books, you can annotate photos, screenshots, PDFs, and more using built-in drawing tools.

Show, move, and hide the Markup toolbar

To show the Markup toolbar in a supported app, tap the Markup Switch button off or Markup, then do any of the following:

- Move the Markup toolbar: Drag the toolbar to any edge of the screen.

(Drag from the middle edge of the toolbar closest to the center of the screen.)

- Automatically minimize the toolbar when you're drawing or entering text: Tap the Ellipsis button ⦿, then turn on Auto-minimize.

To show the full toolbar again, tap the minimized version.

- Hide the toolbar: Tap the Markup Switch button on Ⓐ or Done.

Draw with Markup

In the Markup toolbar, tap the pen, marker, or pencil tool, then write or draw with your finger or Apple Pencil (supported models).

Note: If you don't see the Markup toolbar on supported app, tap the Markup Switch button off or Markup. If the toolbar is minimized, tap its minimized version.

While drawing, do any of the following:

- Change the line weight: Tap the drawing tool in the toolbar, then choose an option.

- Change the opacity: Tap the drawing tool in the toolbar, then drag the slider.

- Change the color: Choose a color from the color picker in the toolbar.

- Undo a mistake: Tap the Undo button.

- Draw a straight line: Tap the ruler tool in the toolbar, then draw a line along the edge of the ruler.

- To change the angle of the ruler, touch and hold the ruler with two fingers, then rotate your fingers.

- To move the ruler without changing its angle, drag it with one finger.

- To make the ruler disappear, tap the ruler tool again.

Erase a mistake

Tap the eraser tool in the Markup toolbar in a supported app, then do one of the following:

- Erase with the pixel eraser: Scrub over the mistake with your finger or Apple Pencil.

- Erase with the object eraser: Touch the object with your finger or Apple Pencil.

- Switch between the pixel and the object erasers: Tap the eraser tool again, then choose Pixel Eraser or Object Eraser.

Note: If you don't see the Markup toolbar, tap the Markup Switch button off Ⓐ or Markup. If the toolbar is minimized, tap its minimized version.

- Move elements of your drawing

- In the Markup toolbar, tap the lasso tool (between the eraser and ruler tools), then drag around the elements to make a selection.

Note: If you don't see the Markup toolbar in a supported app, tap the Markup Switch button off or Markup. If the toolbar is minimized, tap its minimized version.

Move elements of your drawing

1. In the Markup toolbar, tap the lasso tool (between the eraser and ruler tools), then drag around the elements to make a selection.

2. Lift your finger or Apple Pencil, then drag your selection to a new location.

Tip: You can take a screenshot and immediately begin marking it up with Apple Pencil by swiping up from the bottom-left corner of the screen. To mark up a screenshot right after you take it if you don't have Apple Pencil, tap the thumbnail that appears for a few moments in the bottom-left corner of the screen.

Add text, shapes, and signatures with Markup on iPad

In supported apps, you can use Markup to add text, speech bubbles, shapes, and signatures.

Add text

1. In the Markup toolbar in a supported app, tap the Add Annotation button ⊕, then tap Text.

   Note: If you don't see the Markup toolbar, tap the Markup Switch button off Ⓐ or Markup. If the toolbar is minimized, tap its minimized version.

2. Double-tap the text box.

3. Use the keyboard to enter text.

To change text after you add it, tap the text to select it, then do any of the following:

- Change the font, size, or layout: Tap the Shape Attributes button AA in the toolbar, then choose an option.

- Delete, edit, or duplicate the text: Tap Edit, then choose an option.

- Move the text: Drag it.

To hide the Markup toolbar when you finish, tap the Markup Switch button on Ⓐ or Done.

Add a shape

1. In the Markup toolbar in a supported app, tap the Add Annotation button , then choose a shape.

   Note: If you don't see the Markup toolbar, tap the Markup Switch button off or Markup. If the toolbar is minimized, tap its minimized version.

To adjust the shape, do any of the following:

- Move the shape: Drag it.

- Resize the shape: Drag any blue dot along the shape's outline.

- Change the outline color: Tap a color in the color picker.

- Fill the shape with color or change the line thickness: Tap the Shapes Attribute button 🔲, then choose an option.

- Adjust the form of an arrow or speech bubble shape: Drag a green dot.

- Delete or duplicate a shape: Tap it, then choose an option.

- To hide the Markup toolbar when you finish, tap the Markup Switch button on or **Done**.

Add your signature

In the Markup toolbar in a supported app, tap the Add Annotation button, then choose Signature.

Note: If you don't see the Markup toolbar, tap the Markup Switch button off or Markup. If the toolbar is minimized, tap its minimized version.

To hide the Markup toolbar when you finish, tap the Markup Switch button on or Done.

Zoom in or magnify in Markup on iPad

In Markup on supported apps, zoom in to draw the details. Use the magnifier when you only need to see the details.

Zoom in

While using Markup in a supported app, pinch open so you can draw, adjust shapes, and more, up close. To pan when you're zoomed in, drag two fingers. To zoom back out, pinch closed.

Magnify

In the Markup toolbar in a supported app, tap the Add Annotation button, then tap Magnifier.

Note: If you don't see the Markup toolbar, tap the Markup Switch button off or Markup. If the toolbar is minimized, tap its minimized version.

To change the magnifier's characteristics, do any of the following:

- Change the magnification level: Drag the green dot on the magnifier.

- Change the size of the magnifier: Drag the blue dot on the magnifier.

- Move the magnifier: Drag it.

- Change the outline thickness of the magnifier: Tap the Shapes Attribute button 📑, then choose an option.

- Change the outline color of the magnifier: Choose an option from the color picker.

- Remove or duplicate the magnifier: Tap its outline, then tap Delete or Duplicate.

To hide the Markup toolbar when you finish, tap the Markup Switch button on or **Done**.

Install and manage app extensions on iPad

Some apps let you extend the functionality of your iPad. An app extension may appear as a sharing option, an action option, a widget in Today View, a file provider, or a custom keyboard.

App extensions can also help you edit a photo or video in your Photos app. For example, you can download a photo-related app to apply filters to photos.

Download and install app extensions

1. Download the app from the App Store.

2. Open the app, then follow the onscreen instructions.

Manage sharing or action options

1. Tap the Share button ⬆️, then tap More. (You may need to swipe the options left to reveal More.)

2. Turn the sharing or action options on or off.

3. To reorder the options, touch and drag the Reorder button ≡.

4. Tap Done.

Open two items in Split View on iPad

Open two different apps, or two windows from the same app, by splitting the screen into resizable views. For example, open Messages and Maps at the same time in Split View. Or open two Messages windows in Split View and manage two conversations at the same time.

Open a second item in Split View

1. While using an app, swipe up from the bottom edge and pause to reveal the Dock.

2. Touch and hold an app in the Dock, drag it to the right or left edge of the screen, then lift your finger.

3. If two items are already open in Split View, drag over the item you want to replace.

4. To give both views equal space, drag the divider to the center of the screen.

Drag to resize the split.

Close Split View

Drag the app divider to the left or right edge of the screen, depending on which app you want to close.

Turn Split View into Slide Over

Swipe down from the top of the smaller view.

Open an app in Slide Over on iPad

You can use an app that slides in front of another app or in front of itself. For example, open Messages in Slide Over to carry on a conversation while using Maps.

iPad keeps track of the apps you open in Slide Over so that you can switch between them easily.

Open another app in Slide Over

1. While using an app, swipe up from the bottom edge and pause to reveal the Dock.

2. Touch and hold an app in the Dock, then drag it above the Dock.

If an app is already open in Slide Over, it's replaced by the app you drag from the Dock.

To open a third app in Slide Over when the screen is in Split View (on supported models), drag the app from the Dock to the Split View divider.

Switch between apps in Slide Over

Swipe right along the bottom of the Slide Over window, or do the following:

1. Swipe up from the bottom of the Slide Over window.

2. Swipe right, then tap the window you want to view.

Move the Slide Over window

- Do one of the following:

- Move the Slide Over window to the other side of the screen: Drag from the top of the Slide Over window.

- Remove the Slide Over window: Drag the top of the window off the right edge of the screen.

Turn Slide Over into Split View

Drag the top of the Slide Over window to the lower right or lower left of the screen.

On supported models, you can use Slide Over and Split View simultaneously.

View all of an app's workspaces

You can view all open windows for an app, including those in Split View and Slide Over.

- From an open app: Tap the app icon in the Dock.

- Swipe up from the bottom edge of the screen if you don't see the Dock.

- From the Home screen: Touch and hold an app icon, then choose the Show All Windows quick action.

  For an app in the Dock, swipe up from the bottom edge of the screen if you don't see the Dock.

Note: If you touch and hold an app icon for too long before choosing a quick action, the icons begin to jiggle. Tap Done or press the Home button (models with a Home button), then try again.

Chapter 4
Use iPad with iPhone, iPod touch, Mac, and PC

Share your internet connection from iPad (Wi-Fi + cellular)

You can use Personal Hotspot to share a cellular internet connection from your iPad (Wi-Fi + Cellular models) to other devices. Personal Hotspot is useful when the other devices don't have internet access from a Wi-Fi network. Instant Hotspot allows you to connect your devices to Personal Hotspot without entering a password.

If a nearby iPhone or iPad (Wi-Fi + Cellular models) is sharing its Personal Hotspot, you can use its cellular internet connection on your iPad.

Note: Personal Hotspot is not available with all carriers. Additional fees may apply. The number of devices that can join your Personal Hotspot at one time depends on your carrier and iPad model. Contact your carrier for more information.

Set up Personal Hotspot on iPad

Go to Settings ⚙ > Cellular > Personal Hotspot, then turn on Allow Others to Join.

Note: If you don't see the option for Personal Hotspot, and Cellular Data is turned on in Settings > Cellular, contact your carrier about adding Personal Hotspot to your plan.

You can change the following settings:

- *Change the Wi-Fi password for your Personal Hotspot:* Go to Settings > Cellular > Personal Hotspot > Wi-Fi Password.

- *Change the name of your Personal Hotspot:* Go to Settings > General > About > Name.

- *Turn off Personal Hotspot and disconnect devices:* Go to Settings > Cellular > Personal Hotspot, then turn off Allow Others to Join.

Connect iPhone, iPod touch, or another iPad to your Personal Hotspot

On the other device, go to Settings 🔘 > Wi-Fi, then choose your iPad from the list of available networks.

If asked for a password on the other device, enter the password shown in Settings > Cellular > Personal Hotspot on your iPad.

If your iPad and the other device are set up as follows, then Instant Hotspot connects the devices without requiring a password:

- You're signed in with the same Apple ID on each device.

- Each device has Bluetooth turned on.

- Each device has Wi-Fi turned on.

When a device is connected, a blue band appears at the top of your iPad screen. The Personal Hotspot icon appears in the status bar of the connected device.

Connect a Mac or PC to your Personal Hotspot

Do one of the following:

- *Connect your Mac with Wi-Fi and Instant Hotspot:* On your Mac, use the Wi-Fi status menu in the menu bar to choose your iPad from the list of available networks.

  You must be signed in with the same Apple ID on your Mac and iPad, have Bluetooth turned on, and have Wi-Fi turned on.

  The Wi-Fi status icon in the menu bar changes to the Personal Hotspot icon as long as your Mac remains connected to your Personal Hotspot.

- *Connect a Mac or PC with Bluetooth:* To make sure your iPad is discoverable, go to Settings > Bluetooth and leave the screen showing. Then on your Mac or PC, follow the manufacturer directions to set up a Bluetooth network connection.

- *Connect a Mac or PC with USB:* Connect iPad and your computer using USB. If you see an alert that says Trust this Computer?, tap Trust. In your computer's Network preferences, choose iPad, then configure the network settings.

With Family Sharing, you can share your Personal Hotspot with any member of your family automatically or after they ask for approval

When you share a Personal Hotspot from your iPad, it uses cellular data for the internet connection. To monitor your cellular data network usage, go to Settings > Cellular > Usage.

Make and receive phone calls on iPad

You can make and receive calls on your iPad by relaying calls through your iPhone.

To make calls this way, you must set up FaceTime and sign in with the same Apple ID on both of your devices. (iOS 9, iPadOS 13, OS X 10.10, or later required.)

Note: Wi-Fi Calling on other devices is available with some carriers, and cellular charges may apply.

You must first set up your iPhone, and then set up your iPad.

Allow phone calls on your iPad from your iPhone

1. On your iPhone, go to Settings > Cellular.

2. If your iPhone has Dual SIM, choose a line (below Cellular Plans).

3. Do any of the following:

- Tap Calls on Other Devices, turn on Allow Calls on Other Devices, then choose your iPad along with any other devices on which you'd like to make and receive calls.

   This allows iPad and other devices where you're signed in with the same Apple ID to make and receive calls when they're nearby your iPhone and connected to Wi-Fi.

- Tap Wi-Fi Calling, then turn on Add Wi-Fi Calling For Other Devices.

   This allows iPad and other devices where you're signed in with the same Apple ID to make and receive calls even when your iPhone isn't nearby.

   On your iPad, set up FaceTime and sign in with the same Apple ID that you use on your iPhone.

   Go to Settings > FaceTime, then turn on FaceTime and Calls from iPhone. If you're asked, turn on Wi-Fi calling.

Note: If you enable Wi-Fi Calling, emergency calls may be made over Wi-Fi, and your device's location information may be used for emergency calls to aid response efforts, regardless of whether you enable Location Services. Some carriers may use the address you registered with the carrier when signing up for Wi-Fi Calling as your location.

Make or receive a phone call on your iPad

- *Make a call:* Tap a phone number in Contacts, Calendar, FaceTime, Messages, Spotlight, or Safari. Or open FaceTime, enter a contact or phone number, then tap 📞.

- *Receive a call:* Swipe or tap the notification to answer or ignore the call.

Use iPad as a second display for your Mac

With Sidecar, you can extend the workspace of your Mac by using iPad as a second display. The extended workspace allows you to do the following:

- Use different apps on the different screens.

- Use the same app on both screens. For example, you can view your artwork on your Mac screen while you use Apple Pencil and an app's tools and palettes on iPad.

- Mirror the screens so that Mac and iPad display the same content.

Sidecar requires macOS 10.15 or later and iPadOS 13 or later on supported models.

Use Sidecar

1. Make sure you're signed in with the same Apple ID on your Mac and a nearby iPad.

2. Use one of the following connections:

   - *Wireless:* Make sure your Mac and your iPad have Wi-Fi and Bluetooth turned on. They must also be within Bluetooth range of one another (about 33 feet or 10 meters).

   - *USB:* Connect your Mac and iPad using the appropriate USB cable.

   Click the AirPlay menu ▱ in the menu bar on your Mac, then choose your iPad.

   Do any of the following:

   - *Use the Sidecar menu on Mac:* You can easily change how you work with iPad from the Sidecar menu ▰ in the menu bar. For example, switch between using iPad as a mirrored or separate display, or show or hide the sidebar or Touch Bar on iPad.

- *Move windows from Mac to iPad:* Drag a window to the edge of the screen until the pointer appears on your iPad. Or hold the pointer over the green button in the top-left corner of the window, then choose Move to [*iPad name*].

- *Move windows from iPad to Mac:* Drag a window to the edge of the screen until the pointer appears on your Mac. Or hold the pointer over the green button in the top-left corner of the window, then choose Move Window Back to Mac.

- *Use the sidebar on iPad:* With your finger or Apple Pencil, tap icons in the sidebar to show or hide the menu bar ⬆️, the Dock ⬇️, or the keyboard ⌨️. Or tap one or more modifier keys, such as Ctrl ⌃, to use keyboard shortcuts.

- *Use the Touch Bar on iPad:* With your finger or Apple Pencil, tap any button in the Touch Bar. The buttons available vary depending on the app or task.

- *Use Apple Pencil on iPad:* With your Apple Pencil, tap to select items such as menu commands, checkboxes, or files.

  If you turn on "Enable double tap on Apple Pencil" in Sidecar preferences on your Mac, you can double-tap the lower section of your Apple Pencil (2nd generation) to switch drawing tools in some apps.

- *Use standard gestures on iPad:* Use your fingers to tap, touch and hold, swipe, scroll, and zoom.

- *On iPad, switch between the Mac desktop and the iPad Home Screen:* To show the Home Screen, swipe up from the bottom edge of your iPad. To return to the Mac desktop, tap the Sidecar icon in the Dock on your iPad.

When you're ready to stop using your iPad, tap the Disconnect icon at the bottom of the sidebar on iPad.

You can also disconnect from the Sidecar menu in the menu bar and in Sidecar preferences and Displays preferences on your Mac.

Change Sidecar preferences

1. On your Mac, choose Apple menu > System Preferences, then click Sidecar.

2. Choose from the following options:

- *Show, move, or hide the sidebar on your iPad:* To show the sidebar, select Show Sidebar, then to move it, click the pop-up menu and choose a location. To hide the sidebar, deselect Show Sidebar.

- *Show, move, or hide the Touch Bar on your iPad:* To show the Touch Bar, select Show Touch Bar, then to move it, click the pop-up menu and choose a location. To hide the Touch Bar, deselect Show Touch Bar.

  When you use an app that supports the Touch Bar on your iPad, the Touch Bar is shown in the location you specified. The buttons available in the Touch Bar vary depending on the current app and task.

- *Enable double tap on Apple Pencil:* Select this option to be able to double-tap the lower section of Apple Pencil (2nd generation) to switch drawing tools in some apps.

- *Choose which iPad to connect to:* If you have more than one available iPad, click the "Connect to" pop-up menu, then choose the iPad you want.

Hand off tasks between iPad and your Mac

Continue working on one device where you left off on another. You can use Handoff with many Apple apps—for example, Mail, Safari, Pages, Numbers, Keynote, Maps, Messages, Reminders, Calendar, and Contacts—and even some third-party apps. To use Handoff, you must be signed in with the same Apple ID on all your devices. Your devices must have Bluetooth turned on in Settings and be within Bluetooth range of one another (about 33 feet or 10 meters).

Switch devices

- *From Mac to iPad:* The Handoff icon of the app you're using on your Mac appears on iPad on the right side of the Dock. Tap the Handoff icon to continue working in the app on iPad.

- *From iPad to Mac:* The Handoff icon of the app you're using on iPad appears on your Mac at the left end of the Dock (or the top, depending on the Dock position). Click the icon to continue working in the app.

Disable Handoff on your devices

- *iPad, iPhone, and iPod touch:* Go to Settings 🔘 , then tap General > AirPlay & Handoff.

- *Mac:* Choose Apple Menu > System Preferences > General, then turn off "Allow Handoff between this Mac and your iCloud devices."

Cut, copy, and paste between iPad and your Mac

You can cut or copy content (a block of text or an image, for example) on your iPad, then paste it on another iPhone, iPad, iPod touch, or a Mac computer, and vice versa.

For Universal Clipboard to work, you must be signed in with the same Apple ID on all your devices. Your devices must be connected to Wi-Fi, be within Bluetooth range of one another (about 33 feet or 10 meters), have Bluetooth turned on in

Settings 🔘 , and have Handoff enabled. (iPadOS 13, iOS 10, macOS 10.12, or later required.)

You must cut, copy, and paste your content within a short period of time.

Copy, cut, or paste

- *Copy:* Pinch closed with three fingers.

- *Cut:* Pinch closed with three fingers two times.

- *Paste:* Pinch open with three fingers.

You can also touch and hold a selection, then tap Cut, Copy, or Paste.

Connect iPad and your computer using USB

Using USB, you can directly connect iPad and a Mac or Windows PC to set up your iPad, charge the iPad battery, share your iPad internet connection, transfer files, and sync content.

1. Make sure you have one of the following:

   - Mac with a USB port and OS X 10.9 or later

   - PC with a USB port and Windows 7 or later

   Connect iPad to the USB port on your computer using an appropriate cable.

Depending on the type of USB port on your computer, the cable included with your iPad may be appropriate. Alternatively, you may need one of the following (sold separately):

- A USB-C to Lightning Cable

- A USB-C to USB Adapter, a USB-C Digital AV Multiport Adapter, or a USB-C VGA Multiport Adapter

Sync iPad with your computer

You can use iCloud to automatically keep your photos, files, calendar, and more updated across all your devices where you're signed in with your Apple ID. (You can even use a Windows PC to access your iCloud data on iCloud.com.) Other services like Apple Music allow you to access additional content across your devices. With iCloud and services like Apple Music, no syncing is required.

If you don't want to use iCloud or other services, you can connect iPad to your Mac or Windows PC to sync the following items:

- Albums, songs, playlists, movies, TV shows, podcasts, books, and audiobooks

- Photos and videos

- Contacts and calendars

With syncing, you can keep these items up to date between your computer and your iPad.

Note: If you use iCloud or other services like Apple Music, options for syncing with your computer might not be available.

Set up syncing between your Mac and iPad

1. Connect iPad and your computer using USB.

2. In the Finder sidebar on your Mac, select your iPad.

   Note: To use the Finder to sync content, macOS 10.15 or later is required. With earlier versions of macOS, use iTunes to sync with your Mac.

3. At the top of the window, click the type of content you want to sync (for example, Movies or Books).

4. Select "Sync [*content type*] onto [*device name*]."

By default, all items of a content type are synced, but you can choose to sync individual items, such as selected music, movies, books, or calendars.

5. Repeat steps 3 and 4 for each type of content you want to sync, then click Apply.

Your Mac syncs to your iPad whenever you connect them.

To view or change syncing options, select your iPad in the Finder sidebar, then choose from the options at the top of the window.

Before disconnecting your iPad from your Mac, click the Eject button in the Finder sidebar.

See Sync content between your Mac and iPhone or iPad in the macOS User Guide.

Set up syncing between your Windows PC and iPad

1. Connect iPad and your computer using USB.

2. In the iTunes app on your PC, click the iPad button near the top left of the iTunes window.

3. Select the type of content you want to sync (for example, Movies or Books) in the sidebar on the left.

4. Select Sync to turn on syncing for that type of item.

By default, all items of a content type are synced, but you can choose to sync individual items, such as selected music, movies, books, or calendars.

5. Repeat steps 3 and 4 for each type of content you want to include on your iPad, then click Apply.

By default, your Windows PC syncs to your iPad whenever you connect them. You can have iTunes ask you before syncing, and if there are some items you never want sync, you can keep them from being synced.

Turn on Wi-Fi syncing

1. Connect iPad and your computer using USB.

2. Do one of the following:

- *In the Finder sidebar on your Mac:* Select your iPad, click General at the top of the window, then select "Show this [*device*] when on Wi-Fi."

  Note: To use the Finder to turn on Wi-Fi syncing, macOS 10.15 or later is required. With earlier versions of macOS, use iTunes to turn on Wi-Fi syncing.

- *In the iTunes app on a Windows PC:* Click the iPad button near the top left of the iTunes window, click Summary, then select "Sync with this [*device*] over Wi-Fi" (in Options).

  Click Apply.

By default, whenever iPad is plugged into power and is connected over Wi-Fi to your Mac or to iTunes on your Windows PC, the computer syncs your selected content to iPad.

WARNING: If you delete a synced item from your computer, the item is also deleted from your iPad the next time you sync.

Transfer files between iPad and your computer

You can use iCloud Drive to keep your files up to date and accessible on all your devices, including Windows PCs. You can also transfer files between iPad and other devices by using AirDrop and sending email attachments.

Alternatively, you can transfer files for apps that support file sharing by connecting iPad to a Mac (with a USB port and OS X 10.9 or later) or a Windows PC (with a USB port and Windows 7 or later).

Transfer files between iPad and your Mac

1. Connect iPad to your Mac.

   You can connect using USB, or if you set up Wi-Fi syncing, you can use a Wi-Fi connection.

2. In the Finder sidebar on your Mac, select your iPad.

   Note: To use the Finder to transfer files, macOS 10.15 or later is required. With earlier versions of macOS, use iTunes to transfer files.

3. At the top of the Finder window, click Files, then do one of the following:

- *Transfer from Mac to iPad:* Drag a file or a selection of files from a Finder window onto an app name in the list.

- *Transfer from iPad to Mac:* Click the disclosure triangle beside an app name to see its files on your iPad, then drag a file to a Finder window.

To delete a file from iPad, select it below an app name, press Command-Delete, then click Delete.

Transfer files between iPad and your Windows PC

1. Connect iPad to your Windows PC.

   You can connect using USB, or if you set up Wi-Fi syncing, you can use a Wi-Fi connection.

2. In iTunes on your Windows PC, click the iPad button near the top left of the iTunes window.

3. Click File Sharing, select an app in the list, then do one of the following:

- *Transfer a file from your iPad to your computer:* Select the file you want to transfer in the list on the right, click "Save to," select where you want to save the file, then click Save To.

- *Transfer a file from your computer to your iPad:* Click Add, select the file you want to transfer, then click Add.

To delete a file from iPad, select the file, press the Delete key, then click Delete.

File transfers occur immediately. To view items transferred to iPad, go to On My iPad in the Files app on iPad.

Important: Syncing has no effect on file transfers, so syncing doesn't keep transferred files on iPad up to date with the files on your computer.

Chapter 5

Understanding Various Apps

Find and buy books and audiobooks in Apple Books on iPad

In the Books app 📖, you can find today's bestsellers, view top charts, or browse lists curated by Apple Books editors. After you select a book or audiobook, you can read or listen to it right in the app.

1. Open Books, then tap Book Store or Audiobooks to browse titles, or tap Search to look for a specific title or author.

2. Tap a book cover to see more details, read a sample, listen to a preview, or mark as Want to Read.

3. Tap Buy to purchase a title, or tap Get to download a free title.

   All purchases are made with the payment method associated with your Apple ID.

On iPad models that connect to a cellular network, you can allow books and audiobooks to be downloaded automatically over your cellular network when you aren't connected to Wi-Fi.

Go to Settings ⚙️ > Books, scroll to Cellular Data, tap Downloads, then tap Always Allow.

Read books in the Books app on iPad

In the Books app [book icon], use the Reading Now and Library tabs at the bottom of the screen to see the books you're reading, the books you want to read, your book collections, and more.

- *Reading Now:* Tap to access the books and audiobooks you're currently reading. Scroll down to see books and audiobooks you've added to your Want To Read collection and books you've sampled. You can also set daily reading goals and keep track of the books you finish throughout the year.

- *Library:* Tap to see all of the books, audiobooks, series, and PDFs you got from the Book Store or manually added to your library. You can tap Collections to view books sorted into collections, such as Want to Read, My Samples, Audiobooks, and Finished.

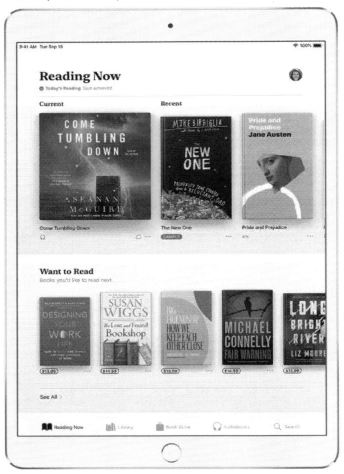

Read a book

Tap the Reading Now or Library tab, then tap a cover to open a book. Use gestures and controls to navigate as follows:

- *Turn the page:* Tap the right side of the page or swipe right to left.

- *Go back to the previous page:* Tap the left side of the page or swipe left to right.

- *Go to a specific page:* Tap the page and move the slider at the bottom of the screen left or right. Or, tap 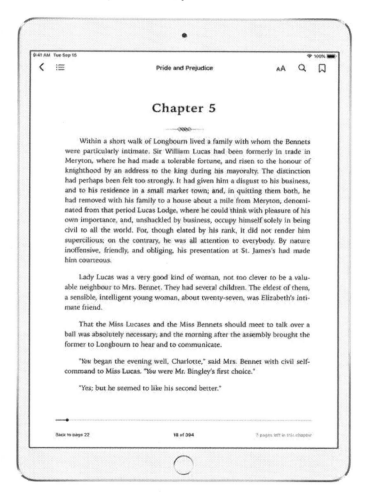 and enter a page number, then tap the page number in the search results.

- *Close a book:* Tap the center of the page to show the controls, then tap $\langle$.

Tip: Turn iPad to landscape orientation to view two pages at once.

When you finish a book, personalized recommendations appear to help you discover your next read.

Change text and display appearance

Tap the page, tap A**A**, then do any of the following:

- *Adjust the screen brightness:* Drag the slider left or right.

- *Change the font size:* Tap the large A to increase the font size or tap the small A to decrease it.

- *Change the font:* Tap Fonts to choose a different font.

- *Change the page background color:* Tap a colored circle.

- *Dim th screen when it's dark:* Turn on Auto-Night Theme to automatically change the page color and brightness when using Books in low-light conditions. (Not all books support Auto-Night Theme.)

- *Turn off pagination:* Turn Vertical Scrolling on to scroll continuously through a book or PDF.

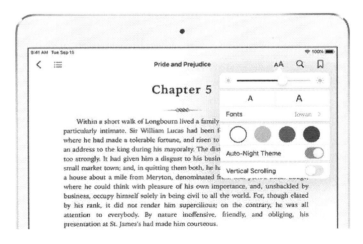

Bookmark a page

When you close a book, your place is saved automatically—you don't need to add a bookmark. Bookmark pages you want to return to again.

Tap ⬓ to add a bookmark; tap it again to remove the Bookmark.

To see all your bookmarks, tap ≔, then tap Bookmarks.

Highlight or underline text

1. Touch and hold a word, then move the grab points to adjust the selection.

2. Tap Highlight, then tap ⬯ to choose a highlight color or underline.

   To remove a highlight or underline, tap the text, then tap 🗑

   .

To see all of your highlights, tap ≡, then tap Notes.

Add a note

1. Touch and hold a word, then move the grab points to adjust the selection.

2. Tap Note, then enter note text.

3. Tap the page to close the note and continue reading.

To see all of your notes, tap :≡, then tap Notes. Swipe left on a note to delete it.

Share a selection

You can send text selections using AirDrop, Mail, or Messages, or you can add the selection to Notes. If the book is from the Book Store, a link to the book is included with the selection. (Sharing may not be available in all countries or regions.)

1. Touch and hold a word, then move the grab points to adjust the selection.

2. Tap Share, then choose a method.

You can also send a link to view the book in the Book Store. Tap a page, tap :≡, then tap ⬆️.

Access your books on all your devices

To keep your Books information updated across your iPhone, iPad, and iPod touch, sign in with the same Apple ID on each device, then do the following:

- *Sync Reading position, bookmarks, notes, and highlights:* Go to Settings ⚙ > [*your name*] > iCloud, then turn on both iCloud Drive and Books.

- *Sync Reading Now, Library, and collections:* Go to Settings > [*your name*] > iCloud, then turn on both iCloud Drive and Books. Then go to Settings > Books, and turn on Reading Now.

Access your books on your Mac

To see your books, audiobooks, and PDFs on your Mac, choose

Apple menu > System Preferences, then do one of the

following:

- *macOS 10.15 or later:* Click Apple ID, select iCloud in the
  sidebar, then select iCloud Drive. Click Options, then select
  Books.

- *macOS 10.14 or earlier:* Click iCloud, then select
  iCloud Drive. Click Options, then select Books.

To see your collections, bookmarks, notes, and highlights on
your Mac, open Books, choose Preferences, click General, then
select "Sync collections, bookmarks, and highlights across
devices."

Listen to audiobooks in Books on iPad

Use the Books app to listen to audiobooks on your iPad.

Play an audiobook

In Reading Now or in the Audiobooks collection in your Library, tap the cover, then do any of the following:

- *Skip forward or back:* Touch and hold the rounded arrows, slide and hold the book cover, or use external controls such as headphones or car controls.

Note: To change the number of seconds that skipping advances, go to Settings ⚙ > Books, then scroll down to Audiobooks.

- *Speed it up, or slow it down:* Tap the playback speed in the lower-left corner to choose a different speed.

- *Set a sleep timer:* Tap 🌙, then choose a duration.

- *Go to a chapter:* Tap ☰, then tap a chapter.

  Note: Some audiobooks refer to chapters as *tracks*, or don't define chapters.

- *Go to a specific time:* Drag the playhead, directly below the audiobook cover. The point where you started listening is marked with a gray circle on the timeline. Tap the circle to jump back to that spot.

If a Wi-Fi connection to the internet isn't available, audiobooks play over your carrier's cellular network, which may result in additional fees. Audiobooks over 200 MB can't be streamed over cellular (iPadOS 13.3 or earlier).

Set reading goals in Books on iPad

The Books app 📖 helps you keep track of how many minutes you read every day, and how many books and audiobooks you finish each year. You can customize your goals to spend more time reading, set new reading streaks, and share your achievements with friends.

Change your daily reading goal

You can adjust your daily reading goal depending on how many minutes you want to read per day. If you don't customize your daily reading goal, it's set to five minutes per day.

1. Tap the Reading Now tab, then swipe down to Reading Goals.

2. Tap Today's Reading, then tap Adjust Goal.

3. Slide the counter up or down to set the minutes per day that you want to read.

When you reach your daily reading goal, you receive a notification from Books; tap it to get more details about your achievement, or send your achievement to friends.

Note: To count PDFs toward your reading goal, go to Settings ⚙ > Books, then turn on Include PDFs.

Change your yearly reading goal

After you finish reading a book or audiobook in Books, the Books Read This Year collection appears below Reading Goals. The default yearly reading goal is three books per year, but you can increase or decrease your goal depending on how many books you want to finish.

1. Tap the Reading Now tab, then swipe down to Books Read This Year.

2. Tap a gray placeholder square, or a book cover, then tap Adjust Goal.

3. Slide the counter up or down to set the books per year that you want to read.

When you reach your yearly reading goal, you receive a notification from Books; tap it to get more details about your achievement, or send your achievement to friends.

See your reading streaks and records

Books lets you know how many days in a row you reach your daily reading goal and notifies you when you set a record.

To view your current reading streak and record, tap the Reading Now tab, then swipe down to Reading Goals.

Turn off notifications and Reading Goals

Turn off notifications: To stop receiving notifications when you achieve a reading goal or set a reading streak, tap your account in the top-right corner of the Reading Now tab, tap Notifications, then turn off Reading Goals.

Turn off Reading Goals: Go to Settings > Books > then turn off Reading Goals. When Reading Goals is turned off, the reading indicators in Reading Now are hidden and you don't receive notifications.

Organize books in the Books app on iPad

In the Books app , the books and audiobooks you purchase are saved in your library and automatically sorted into collections, such as Audiobooks, Want to Read, and Finished.

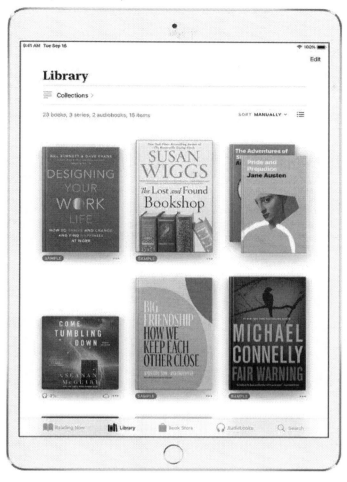

Create a collection and add books to it

You can create your own collections to personalize your library.

1. Tap Library, tap Collections, then tap New Collection.

2. Name the collection, for example, *Beach Reads* or *Book Club*, then tap Done.

3. To add a book to the collection, tap ●●● below the book cover (or on the book's details page in the Book Store), tap Add to Collection, then choose the collection.

You can add the same book to multiple collections.

Sort books in your library

Choose how the books in your library are sorted and appear.

1. Tap Library, then scroll down and tap the word that appears next to Sort or Sort By.

2. Choose Recent, Title, Author, or Manually.

   If you choose Manually, touch and hold a book cover, then drag it to the position you want.

3. Tap ⦂☰ to view books by title or cover.

Tip: You can sort books in a collection the same way.

Remove books, audiobooks, and PDFs

You can remove books, audiobooks, and PDFs from Reading Now and your library collections, or hide them on your iPad.

1. Tap Library, then tap Edit.

   For audiobooks, tap Library, then go to the Audiobooks collection.

2. Tap the items you want to remove.

3. Tap 🗑 and select an option.

To unhide books that you have hidden, tap Reading Now, tap your account icon, then tap Manage Hidden Purchases.

Access your library across devices

You can sync your Library and collections across all your devices where you are signed in with the same Apple ID. Go to Settings ⚙️ > [*your name*] > iCloud, turn on iCloud Drive, then turn on Books.

Read PDF documents in Books on iPad

In the Books app 📖, you can open and save PDFs that you receive in Mail, Messages, and other apps.

Open PDFs in Books

Tap the PDF attachment to open it, tap 📤, then tap Books.

Share or print a PDF document

Open the PDF document, tap 📤, then choose a share option such as AirDrop, Mail, or Messages, or tap Print.

Mark up a PDF

Open the PDF and tap Ⓐ to use the drawing and annotation tools (tap near the center of a page if you don't see Ⓐ).

See Draw in apps with Markup on iPad.

View PDFs across devices

You can see PDFs and books that are not from the Book Store across your iPhone, iPad, iPod touch, and Mac where you're signed in with the same Apple ID.

Go to Settings  > [your name] > iCloud, turn on iCloud Drive, then turn on Books. Then go to Settings > Books, and turn on iCloud Drive.

Create and edit events in Calendar on iPad

Use the Calendar app to create and edit events, appointments, and meetings.

View invitations.

Change calendars
or accounts.

Ask Siri. Say something like:

- "Set up a meeting with Gordon at 9"

- "Do I have a meeting at 10?"

- "Where is my 3:30 meeting?"

Add an event

1. In day view, tap ✛ near the top left.

2. Fill in the event details.

   Enter the name and location of the event, the start and end times, how often it repeats, and so on.

3. Tap Add.

Add an alert

You can set an alert to be reminded of an event beforehand.

1. Tap the event, then tap Edit near the top right.

2. In the event details, tap Alert.

3. Choose when you want to be reminded.

   For example, "At time of event," "5 minutes before," or another choice.

   Note: If you add the address of the event's location, Calendar uses Apple Maps to look up locations, traffic conditions, and transit options to tell you when it's time to leave.

Add an attachment

You can add an attachment to a Calendar event to share with invitees.

1. Tap the event, then tap Edit near the top right.

2. In the event details, tap Add attachment.

   The Files app opens, displaying your recently opened files.

3. Locate the file you want to attach.

   To find the file, you can enter its name in the search field, scroll, tap folders to open them, tap Browse to look in other locations (such as iCloud Drive), and so on.

   4. Tap Done.

To remove the attachment, tap the event, tap Edit near the top right, swipe left over the attachment, then tap Remove.

Find events in other apps

Siri can suggest events found in Mail, Messages, and Safari—such as flight reservations and hotel bookings—so you can add them easily in Calendar.

1. Go to Settings ⚙ > Calendar > Siri & Search.

2. Turn on Show Siri Suggestions in App to allow Siri to suggest events found in other apps.

To allow Siri to make suggestions in other apps based on how you use Calendar, turn on Learn from this App.

Edit an event

You can change the time of an event and any of the other event details.

- *Change the time:* In day view, touch and hold the event, then drag it to a new time, or adjust the grab points.

- *Change event details:* Tap the event, tap Edit near the top right, then in the event details, tap a setting to change it, or tap in a field to type new information.

Delete an event

In day view, tap the event, then tap Delete Event at the bottom of the screen.

Send and receive invitations in Calendar on iPad

In the Calendar app [15], send and receive meeting and event invitations. iCloud, Microsoft Exchange, and some CalDAV servers let you send and receive meeting invitations. (Not all calendar servers support every feature.)

Invite others to an event

You can invite people to an event, even if you're not the one who scheduled it, with Exchange and some other servers.

1. Tap the event, tap Edit, then tap Invitees.

Or, if you didn't schedule the event, tap it, tap Invitees, then

tap .

2. Type the names or email addresses of invitees, or tap ⊕ to browse your Contacts.

3. Tap Done (or tap Send if you didn't schedule the event).

With Microsoft Exchange, and some other servers, you can invite people to an event even if you're not the one who scheduled it.

If you don't want to be notified when someone declines a meeting, go to Settings 🔘 > Calendar, then turn off Show Invitee Declines.

Reply to an event invitation

1. To respond to an event notification, tap it.

Or, in Calendar, tap ▭, then tap an invitation.

2. Tap your response—Accept, Maybe, or Decline.

To respond to an invitation you receive by email, tap the underlined text in the email, then tap Show in Calendar.

If you add comments to your response (comments may not be available for all calendars), your comments can be seen by the organizer but not by other attendees. To see events you declined, tap ▦, then turn on Show Declined Events.

Schedule a meeting without blocking your schedule

You can add an event to your calendar without having the timeframe appear as busy to others who send you invitations.

1. Tap the event, then tap Edit.

2. Tap Show As, then tap Free.

Suggest a different meeting time

You can suggest a different time for a meeting invitation you've received.

1. Tap the meeting, then tap Propose New Time.

2. Tap the time, then enter a new one.

3. Tap Done, then tap Send.

Quickly send an email to attendees

1. Tap an event that has attendees.

2. Tap Invitees, then tap ✉ .

Change how you view events in Calendar on iPad

In the Calendar app 📅, you can view one day, a week, a month, or a year at a time, or view a list of upcoming events. To change your view of Calendar, do any of the following:

- *Zoom in or out:* Tap Day, Week, Month, or Year at the top of the screen to zoom in or out on your calendar. In week or day view, pinch to zoom in or out.

- *View upcoming events:* Tap to view upcoming events as a list.

Take photos with your iPad camera

Learn how to take great photos with Camera on your iPad. Choose from camera modes such as Photo, Pano, and Square, and use camera features such as Burst and Live Photos.

Ask Siri. Say something like: "Open Camera

Take a photo

Photo is the standard mode that you see when you open Camera. Use Photo mode to take still photos. Swipe the mode selector up or down to choose a different mode, such as Video, Pano, Time-lapse, Slo-mo, and Portrait (on supported models).

1. Tap 📷 on the Home screen or swipe left on the Lock screen to open Camera in Photo mode.

2. Tap the Shutter button or press either volume button to take the photo.

To turn the flash on or off on models that support True Tone Flash or Retina Flash, tap ⚡, then choose Auto, On, or Off.

To set a timer, stabilize your iPad and frame your shot. Tap ⏱, then tap 3s or 10s.

Note: For your security, a green dot appears at the top of the screen when Camera is in use.

Zoom in or out

- On all models, open Camera and pinch the screen to zoom in or out.

- On iPad Pro 11-inch (2nd generation) and iPad Pro 12.9-inch (4th generation), tap 1x on the left side of the screen to zoom out and toggle between 1x and 0.5x. To zoom in, touch and hold 1x, then drag the slider up.

- On all other models, drag the slider on the left side of the screen up or down.

Take a panorama photo

1. Choose Pano mode, then tap the Shutter button.

2. Pan slowly in the direction of the arrow, keeping it on the center line.

3. To finish, tap the Shutter button again.

113

Tap the arrow to pan in the opposite direction. To pan vertically, rotate iPad to landscape orientation. You can reverse the direction of a vertical pan, too.

Take a selfie

1. On iPad Pro 11-inch (2nd generation) and iPad Pro 12.9-inch (4th generation), tap 🔄 to switch to the front camera. On all other models, tap 🔄.

2. Hold your iPad in front of you.

3. Tap the Shutter button or press either volume button to take the shot.

To take a mirrored selfie that captures the shot as you see it in the camera frame, go to Settings ⚙️ > Camera, then turn on Mirror Front Camera (on supported models).

Take a selfie in Portrait mode

On supported models, you can apply a depth-of-field effect to your selfies with the front camera. This effect keeps your face sharp while creating a beautifully blurred background.

1. Choose Portrait mode.

   The front camera is now active.

2. Frame yourself in the yellow portrait box.

3. Tap the Shutter button to take the shot.

Adjust Portrait Lighting in Portrait mode selfies

On models that support Portrait Lighting, you can apply studio-quality lighting effects to your Portrait mode selfies.

1. Choose Portrait mode, then frame your selfie.

2. Drag ⬡ to choose a lighting effect:

   - *Natural Light:* The face is in sharp focus against a blurred background.

   - *Studio Light:* The face is brightly lit, and the photo has an overall clean look.

   - *Contour Light:* The face has dramatic shadows with highlights and lowlights.

   - *Stage Light:* The face is spotlit against a deep black background.

   - *Stage Light Mono:* The effect is similar to Stage Light, but the photo is in classic black and white.

   - *High-Key Light Mono:* Creates a grayscale subject on a white background—iPad Pro 11-inch (2nd generation) and iPad Pro 12.9-inch (4th generation) only.

   Tap the Shutter button to take the shot.

Adjust Depth Control in Portrait mode selfies

On models that support Depth Control, use the Depth Control slider to adjust the level of background blur in your Portrait mode selfies.

1. Choose Portrait mode, then frame your selfie.

2. Tap *f* on the right side of the screen.

   The Depth Control slider appears on the right.

3. Drag the slider up or down down to adjust the effect.

4. Tap the Shutter button to take the shot.

After you take a photo in Portrait mode, you can use the Depth Control slider in Photos to further adjust the background blur effect. See Adjust Depth Control in Portrait mode photos.

Take Burst shots

Burst takes multiple high-speed photos so that you have a range of photos to choose from. You can take Burst photos with the front and rear cameras.

1. Choose Photo or Square mode.

2. Touch and hold the Shutter button to take rapid-fire photos.

   The counter shows how many shots you took.

3. Lift your finger to stop.

4. To select the photos you want to keep, tap the Burst thumbnail, then tap Select.

   Gray dots below the thumbnails mark the suggested photos to keep.

5. Tap the circle in the lower-right corner of each photo you want to save as an individual photo, then tap Done.

To delete the entire group of Burst photos, tap the thumbnail, then tap 🗑.

Take a Live Photo

A Live Photo captures what happens just before and after you take your photo, including the audio.

1. On models that support Live Photos, choose Photo mode.

2. Tap ◎ to turn Live Photos on (yellow is on) or off.

3. Tap the Shutter button to take the shot.

You can edit Live Photos in Photos. In your albums, Live Photos are marked with "Live" in the corner.

Take videos with your iPad camera

Use Camera 📷 to record videos on your iPad and change modes to take slow-motion and time-lapse videos.

Record a video

1. Choose Video mode.

2. Tap the Record button or press either volume button to start recording.

Pinch the screen to zoom in and out.

3. Tap the Record button or press either volume button to stop recording.

By default, video records at 30 fps (frames per second). Depending on your model, you can choose other frame rates and video resolution settings in Settings ⚙ > Camera > Record Video. The faster the frame rate and the higher the resolution, the larger the resulting video file.

Note: For your security, a green dot appears at the top of the screen when Camera is in use

Use quick toggles to change video resolution and frame rate

In Video mode, you can display quick toggles to change the video resolution and frame rates available on your iPad. To display quick toggles in Video mode, go to Settings ⚙ > Camera > Record Video, then turn on Video Format Control.

Record a slow-motion video

1. Choose Slo-mo mode.

2. Tap the Record button or press either volume button to start and stop recording.

To set a portion of the video to play in slow motion and the rest at regular speed, tap the video thumbnail, then tap Edit. Slide the vertical bars below the frame viewer to define the section you want to play back in slow motion.

Depending on your model, you can change the frame rate and resolution. The faster the frame rate and the higher the resolution, the larger the resulting video file.

To change Slo-mo recording settings, go to Settings ⚙ > Camera > Record Slo-mo.

Capture a time-lapse video

1. Choose Time-lapse mode.

2. Set up your iPad where you want to capture a sunset, traffic flowing, or other experience over a period of time.

3. Tap the Record button to start recording; tap it again to stop recording.

On models that support Auto FPS, when you take time-lapse 1080p video at 30 fps under low-light conditions, iPad can automatically reduce the frame rate to 24 fps to improve the video quality. Go to Settings ⚙ > Camera > Record Video, then turn on Auto Low Light FPS.

Use the Camera settings on iPad

Learn how to use the focus, exposure, and other Camera
settings on your iPad.

Adjust the focus and exposure

Before you take a photo, the iPad camera automatically sets the focus and exposure, and face detection balances the exposure across many faces. To manually adjust the focus and exposure, follow these steps:

1. Tap the screen to reveal the automatic focus area and exposure setting.

2. Tap where you want to move the focus area.

3. Next to the focus area, drag ☀ up or down to adjust the exposure.

   To lock your manual focus and exposure settings for upcoming shots, touch and hold the focus area until you see AE/AF Lock; tap the screen to unlock settings.

Align your shots

To display a grid on the camera screen that can help you straighten your shots, go to Settings ⚙ > Camera, then turn on Grid.

Preserve camera settings

You can preserve the last camera mode you used so it's not reset when you next open Camera.

- Go to Settings ⚙️ > Camera > Preserve Settings.

- On supported models, you can also preserve Live Photos settings.

Adjust the shutter sound volume

Adjust the volume of the shutter sound using the volume buttons on the side of your iPad. Or, when Camera is open, swipe down from the top-right corner of the screen to open Control Center, then drag 🔊.

Mute the sound using the volume buttons or the Ring/Silent switch, if your iPad has one. (In some countries or regions, muting is disabled.)

Note: The camera shutter doesn't make a sound when Live Photos ⦿ is turned on.

Adjust HDR camera settings on iPad

HDR (High Dynamic Range) in Camera 📷 helps you get great shots in high-contrast situations. On supported models, the iPad camera takes three photos in rapid succession at different exposures and blends them together. The resulting photo has better detail in the bright and midtone areas.

Take an HDR photo

- On models that take Auto HDR photos and models that take Smart HDR photos, iPad automatically uses HDR when it's most effective.

  To manually control HDR on these models, go to Settings ⚙ > Camera, then turn off Smart HDR. On the camera screen, tap HDR to turn it on or off.

- On all other models, tap HDR on the camera screen to turn on HDR.

By default, the HDR version of the photo is saved in Photos. To save both the HDR and non-HDR version, go to Settings ⚙ > Camera, then turn on Keep Normal Photo.

View, share, and print photos on iPad

All photos and videos you take with Camera 📷 are saved in Photos. With iCloud Photos turned on, all new photos and videos are automatically uploaded and available in Photos on all your devices that are set up with iCloud Photos (with iOS 8.1, iPadOS 13, or later).

Note: If Location Services is turned on in Settings ⊚ > Privacy, photos and videos are tagged with location data that can be used by apps and photo-sharing websites.

View your photos

1. In Camera, tap the thumbnail image below the Shutter button.

2. Swipe right to see the photos you've taken recently.

   Tap the screen to show or hide the controls.

3. Tap All Photos to see all your photos and videos saved in Photos.

Share and print your photos

1. While viewing an image, tap ⬆️.

2. Select an option such as AirDrop, Mail, or Messages to share your photo.

3. Swipe up to select Print from the list of options.

Upload and sync photos across devices

Use iCloud Photos to upload photos and videos from your iPad to iCloud and access them on your iPhone, iPad, or iPod touch where you're signed in using the same Apple ID. iCloud Photos is useful if you want to keep your photos up to date across multiple devices or save space on your iPad. To turn on iCloud

Photos, go to Settings ⚙ > Photos. When iCloud Photos is turned off, you can still collect up to 1000 of your most recent photos in the My Photo Stream album on devices set up with iCloud.

Scan a QR code with the iPad camera

You can use the Camera 📷 to scan Quick Response (QR) codes for links to websites, apps, coupons, tickets, and more. The camera automatically detects and highlights a QR code.

Use the camera to read a QR code

1. Open Camera, then position iPad so that the code appears on the screen.

2. Tap the notification that appears on the screen to go to the relevant website or app.

Open the QR code reader from Control Center

1. Go to Settings ⚙ > Control Center, then tap ➕ next to QR Code Reader.

2. Open Control Center, tap the QR code reader, then position iPad so that the code appears on the screen.

3. To add more light, tap the flashlight to turn it on.

See the time in cities worldwide on iPad

Use the Clock app  to see the local time in different time zones around the world.

Ask Siri. Say something like: "What time is it?" or "What time is it in London?" Learn how to ask Siri.

1. Tap World Clock.

2. To manage your list of cities, tap Edit, then do any of the following:

   - *Add a city:* Tap ╈ , then choose a city.

- *Delete a city:* Tap ⊖ .

- *Reorder the cities:* Touch and hold a clock, then drag it to a new position.

Set an alarm on iPad

In the Clock app 🕐, you can set an alarm that plays a sound at a specific time.

Ask Siri. Say something like: "Wake me up tomorrow at 7 a.m." or "Set an alarm for 9 a.m. every Friday."

Set an alarm

1. Tap Alarm, then tap ➕ .

2. Set the time, then choose any of the following options:

   - *Repeat:* Choose the days of the week.

   - *Label:* Give the alarm a name, like "Water the plants."

   - *Sound:* Choose a tone or a song.

   - *Snooze:* Give yourself nine more minutes.

   Tap Save.

To change or delete the alarm, tap Edit.

Use the timer or stopwatch on iPad

In the Clock app 🕐, you can use the timer to count down from a specified time. You can also use the stopwatch to measure the duration of an event.

Ask Siri. Say something like: "Set the timer for 3 minutes" or "Stop the timer."

Track time with the stopwatch

1. Tap Stopwatch.

   Note: With iPad in portrait orientation, you can switch between the digital and analog faces by swiping the stopwatch.

2. Tap Start.

   The timing continues even if you open another app or if iPad goes to sleep.

3. To record a lap or split, tap Lap.

4. Tap Stop to record the final time.

5. Tap Reset to clear the stopwatch.

Set the timer

1. Tap Timer.

2. Set the duration of time and a sound to play when the timer ends.

Tip: If you want to fall asleep while playing audio or video, you can set the timer to stop the playback. Tap Stop Playing at the bottom of the list.

3. To start the timer, tap Start.

The timer continues even if you open another app or if iPad goes to sleep.

Add and use contact information on iPad

In the Contacts app , you can view and edit your contacts lists from personal, business, and other accounts. You can also create contacts and set up a contact card with your own information.

Ask Siri. Say something like:

- "What's my brother's work address?"

- "Sarah Milos is my sister"

- "Send a message to my sister"

Create a contact

Tap ✛.

Siri also suggests new contacts based on your use of other apps, such as email you receive in Mail and invitations you receive in Calendar. (To turn this feature off, go to

Settings 🔘 > Contacts > Siri & Search, then turn off Show Siri Suggestions for Contacts.)

Based on how you use Contacts, Siri also provides contact information suggestions in other apps. (To turn this feature off, go to Settings 🔘 > Contacts > Siri & Search, then turn off Learn from this App.)

Find a contact

Tap the search field at the top of the contacts list, then enter a name, address, phone number, or other contact information.

Share a contact

Tap a contact, tap Share Contact, then choose a method for sending the contact information.

Sharing the contact sends all of the info from the contact's card.

Quickly reach a contact

To start a message, make a phone call or a FaceTime call, compose an email, or send money with Apple Pay, tap a button below the contact's name.

Send a message.

Make a
FaceTime call.

Open in Maps.

To change the default phone number or email address for a contact method, touch and hold the button for that method below the contact's name, then tap a selection in the list.

Delete a contact

1. Go to the contact's card, then tap Edit.

2. Scroll down, then tap Delete Contact.

Edit contacts on iPad

In the Contacts app , assign a photo to a contact, change a label, add a birthday, and more.

1. Tap a contact, then tap Edit.

2. Do any of the following:

- *Assign a photo to a contact:* Tap "add photo." You can take a photo or add one from the Photos app.

- *Change a label:* Tap the label, then select one in the list, or tap Add Custom Label to create one of your own.

130

- *Add a birthday, social profile, related name, and more:* Tap ⊕ next to the item.

- *Allow calls or texts from a contact to override Do Not Disturb:* Tap Ringtone or Text Tone, then turn on Emergency Bypass.

- *Add notes:* Tap the Notes field.

- *Add a prefix, phonetic name, pronunciation, and more:* Tap "add field," then select an item in the list.

- *Delete contact information:* Tap ⊖ next to a field.

  When you're finished, tap Done.

To change how your contacts are sorted and displayed, go to Settings ⚙ > Contacts.

Set up FaceTime on iPad

In the FaceTime app 📹, you can make video or audio calls to friends and family, whether they're using an iPhone, iPad, iPod touch, or a Mac. With the front camera, you can talk face-to-face; switch to the rear camera to share what you see around you. To capture a moment from your conversation, take a FaceTime Live Photo.

Note: FaceTime, or some FaceTime features, may not be available in all countries or regions.

1. Go to Settings ⚙ > FaceTime, then turn on FaceTime.

2. If you want to be able to take Live Photos during FaceTime calls, turn on FaceTime Live Photos.

3. Enter your phone number, Apple ID, or email address to use with FaceTime.

Make and receive FaceTime calls on iPad

With an internet connection and an Apple ID, you can make and receive calls in the FaceTime app 📹 (first sign in with your Apple ID, or create an Apple ID, if you don't have one). On iPad Wi-Fi + Cellular models, you can also make FaceTime calls over a cellular data connection, which may incur additional charges. To turn this feature off, go to Settings ⚙ > Cellular, then turn off FaceTime.

Drag your image
to any corner.

Tap to take a
Live Photo.

Switch to the
rear camera.

Tap to add stickers
or other fun effects.

Make a FaceTime call

Ask Siri. Say something like: "Make a FaceTime call" or "Call Eliza's mobile." Learn how to ask Siri.

1. In FaceTime, tap ✛ at the top of the screen.

2. Type the name or number you want to call in the entry field at the top, then tap Video ▢◁ to make a video call or tap Audio ✆ to make a FaceTime audio call (not available in all countries or regions).

You can also tap ⊕ to open Contacts and start your call from there, or tap a contact in your list of FaceTime calls to quickly make a call.

Tip: To see more during a FaceTime video call, rotate iPad to use landscape orientation.

Leave a message

If no one answers your FaceTime call, do one of the following:

- Tap Leave a Message.

- Tap Cancel to cancel the call.

- Tap Call Back to try calling back.

Start a FaceTime call from a Messages conversation

In a Messages conversation, you can start a FaceTime call to the person you're chatting with.

1. In the Messages conversation, tap the profile picture, 👤 , or the name at the top of the conversation.

2. Tap FaceTime.

Call again

In your call history, tap the name or number, or tap (i) to choose a name or number in Contacts and start your call from there.

Receive a FaceTime call

When a FaceTime call comes in, tap any of the following:

- *Accept:* Take the call.

- *Decline:* Decline the call.

- *Remind Me:* Set a reminder to call back.

- *Message:* Send a text message to the caller.

Send the caller
a text message.

Set up a reminder
to return a call.

If you're on a regular call when a FaceTime call comes in, instead of *Accept*, you see the *End & Accept* option, which terminates the previous call and connects you to the incoming call.

Delete a call from your call history

In FaceTime, swipe left over the call in your call history, then tap Delete.

Take a Live Photo in FaceTime on iPad

When you're on a video call in the FaceTime app ![camera icon], you can take a FaceTime Live Photo to capture a moment of your conversation (not available in all countries or regions). The camera captures what happens just before and after you take the photo, including the audio, so you can see and hear it later just the way it happened.

To take a FaceTime Live Photo, first make sure FaceTime Live Photos is turned on in Settings ![settings icon] > FaceTime, then do one of the following:

- *On a call with one other person:* Tap ![circle icon].

- *On a Group FaceTime call:* Tap the tile of the person you want to photograph, tap ![expand icon], then tap ![circle icon].

You both receive a notification that the photo was taken, and the Live Photo is saved in your Photos app.

Make a Group FaceTime call on iPad

In the FaceTime app 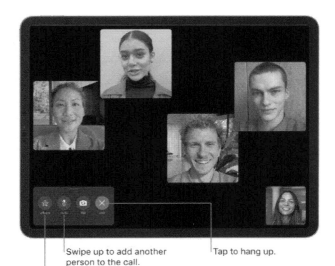, you can invite up to 32 participants to a Group FaceTime call (not available in all countries or regions).

Start a Group FaceTime call

1. In FaceTime, tap ╋ at the top of the screen.

2. Type the names or numbers of the people you want to call in the entry field at the top.

   You can also tap ⊕ to open Contacts and add people from there.

3. Tap Video ⬜◁ to make a video call or tap Audio ✆ to make a FaceTime audio call.

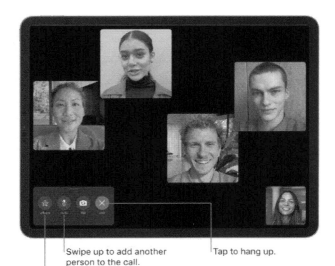

    Swipe up to add another   Tap to hang up.
    person to the call.

Tap to add stickers
or other fun effects.

Each participant appears in a tile on the screen. When a participant speaks (verbally or by using sign language) or you tap the tile, that tile moves to the front and becomes more prominent. Tiles that can't fit on the screen appear in a row at the bottom. To find a participant you don't see, swipe through the row. (The participant's initials may appear in the tile if an image isn't available.)

To prevent the tile of the person speaking from becoming larger during a Group FaceTime call, go to Settings > FaceTime, then turn off Speaking below Automatic Prominence (iPadOS 13.5 or later required).

Note: Sign language detection requires a supported model for the presenter. In addition, both the presenter and participants need iOS 14, iPadOS 14, macOS Big Sur 11.0, or later.

Start a Group FaceTime call from a group Messages conversation

In a group Messages conversation, you can initiate a Group FaceTime call with all the same people you're chatting with in the Messages conversation.

1. In the Messages conversation, tap the names or profile pictures at the top of the conversation.

2. Tap FaceTime.

Add another person to a call

Any participant can add another person at any time during a call.

1. During a FaceTime call, tap the screen to open the controls (if they aren't visible), swipe up from the top of the controls, then tap Add Person.

2. Type the name, Apple ID, or phone number of the person you want to add in the entry field at the top.

   Or tap to add someone from Contacts.

3. Tap Add Person to FaceTime.

Join a Group FaceTime call

When someone invites you to join a Group FaceTime call, you see the incoming call. If you decline the call, you receive a notification that you can tap to join the call at any time while it's active.

Leave a Group FaceTime call

To leave a group call at any time, tap ⊗.

The call remains active if two or more participants remain.

Connect external devices or servers with Files on iPad

You can use the Files app to access files stored on external devices or servers, such as USB drives and SD cards, file servers, and other cloud storage providers like Box and Dropbox, after you connect them to your iPad.

Connect a USB drive or an SD card

1. Insert a USB camera adapter or an SD card reader into the charging port on iPad.

   Depending on your model, use the Lightning to USB Camera Adapter, Lightning to USB 3 Camera Adapter, USB-C to SD Card Camera Reader, or Lightning to SD Card Camera Reader (all sold separately).

   Note: The Lightning to USB 3 Camera Adapter can be powered with a USB power adapter. This allows you to connect USB devices with higher power requirements, such as external hard drives, to iPad.

2. Do one of the following:

   - *Connect a USB drive:* Use the USB cable that came with the USB drive to connect the drive to the camera adapter.

- *Insert an SD memory card into the card reader:* Don't force the card into the slot on the reader; it fits only one way.

  To view the contents of the device, tap Browse at the bottom of the screen, then tap the name of the device below Locations. If you don't see Locations, tap Browse again at the bottom of the screen.

To disconnect the device, simply remove it from the connector on iPad.

Connect to a computer or file server

1. Tap ••• at the top of the Browse sidebar.

   If you don't see the Browse sidebar, tap Browse at the bottom of the screen.

2. Tap Connect to Server.

3. Enter a local hostname or a network address, then tap Connect.

   Tip: After you connect to a computer or file server, it appears in the Recent Servers list on the Connect to Server screen. To connect to a recent server, tap its name.

4. Select how you want to connect:

   - *Guest:* You can connect as a Guest user if the shared computer permits guest access.

- *Registered User:* If you select Registered User, enter your user name and password.

  Tap Next, then select the server volume or shared folder in the Browse sidebar (under Shared).

To disconnect from the file server, tap ⏏ next to the server in the Browse sidebar.

For information on how to set up your Mac to share files, see Set up file sharing on Mac in the macOS User Guide.

Add a cloud storage service

1. Download the app from the App Store, then open the app and follow the onscreen instructions.

2. Open Files, tap More Locations (below Locations in the Browse sidebar), then turn on the service.

3. To view your contents, tap Browse at the bottom of the screen, then tap the name of the storage service below Locations. If you don't see Locations, tap Browse again at the bottom of the screen.

View files and folders in Files on iPad

In the Files app 📁, view and open your documents, images, and other files.

View recently opened files

Tap Recents at the bottom of the screen.

Browse and open files and folders

1. Tap Browse at the bottom of the screen, then tap an item in the Browse sidebar.

   If you don't see the Browse sidebar, tap Browse again.

2. To open a file, location, or folder, tap it.

   Note: If you haven't installed the app that created a file, a preview of the file opens in Quick Look.

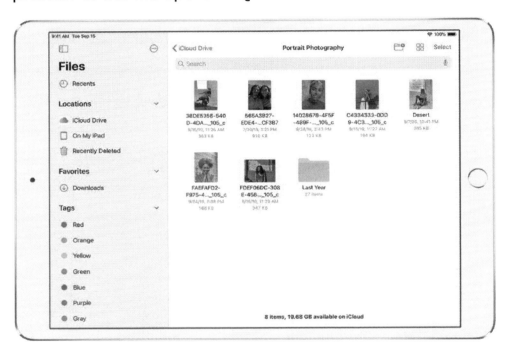

For information about marking folders as favorites or adding tags,

Change how files and folders are sorted

From an open location or folder, drag down from the center of the screen, then tap Name, Date, Size, Kind, or Tags at the top of the screen.

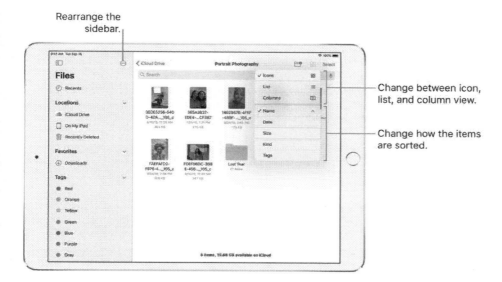

Rearrange the sidebar.

Change between icon, list, and column view.

Change how the items are sorted.

Change to icon, list, or column view

From an open location or folder, drag down from the center of the screen, then do one of the following:

- *View as icons:* Tap ⁝☰ .

- *View as a list:* Tap ⊞ .

- *View as columns:* Tap ⊞ .

  To look deeper into a folder hierarchy from the column view, tap an item in the rightmost column, then swipe left. To see a preview of a file along with its metadata (such as its kind and size), tap the file. (If the preview doesn't appear in the rightmost column, swipe left.) To view the file and perform various actions on it without leaving Files, tap Open under the file preview.

Find a specific file or folder

Enter a filename, folder name, or document type in the search field.

When you search, you have these options:

- *Focus the scope of your search:* Below the search field, tap Recents or the name of the location or tag.

- *Hide the keyboard and see more results on the screen:* Tap ⌨︎ ⌄ .

- *Start a new search:* Tap ⊗ in the search field.

- *Open a result:* Tap it.

Rearrange the Browse sidebar

Tap ••• at the top of the sidebar, tap Edit, then do any of the following:

- *Hide a location:* Turn the location off.

- *Delete a tag and remove it from all items:* Tap ⊖ next to the tag. *Remove an item from the Favorites list:* Tap ⊖ next to the item. (See Mark a folder as a favorite.)

- *Change the order of an item:* Touch and hold ▬, then drag it to a new position.

Share your location in Find My on iPad

Before you can use the Find My app 🔘 to share your location with friends, you need to set up location sharing.

Set up location sharing

1. Tap Me, then turn on Share My Location.

   The device sharing your location appears below My Location.

2. If your iPad isn't currently sharing your location, tap Use This iPad as My Location.

Note: You can share your location from an iPhone, iPad, or iPod touch. To share your location from another device, open Find My on the device and change your location to that device. If the device has iOS 12 or earlier. If you share your location from an iPhone that's paired with Apple Watch (GPS + Cellular models), your location is shared from your Apple Watch when you're out of range of your iPhone and Apple Watch is on your wrist.

You can also change your location sharing settings in Settings 🔘 > [*your name*] > Find My.

Set a label for your location

You can set a label for your current location to make it more meaningful (like Home or Work). When you tap Me, you see the label in addition to your location.

1. Tap Me, then tap Edit Location Name.

2. Select a label.

   To add a new label, tap Add Custom Label, enter a name, then tap Done.

Share your location with a friend

1. Tap People.

2. Scroll to the bottom of the People list, then tap Share My Location.

3. In the To field, type the name of a friend you want to share your location with (or tap ⊕ and select a contact).

4. Tap Send and choose how long you want to share your location.

You can also notify a friend or family member when your location changes.

Stop sharing your location

You can stop sharing your location with a specific friend or hide your location from everyone.

- *Stop sharing with a friend:* Tap People, then tap the name of the person you don't want to share your location with. Tap Stop Sharing My Location, then tap Stop Sharing Location.

- *Hide your location from everyone:* Tap Me, then turn off Share My Location.

Respond to a location sharing request

1. Tap People.

2. Tap Share below the name of the friend who sent the request and choose how long you want to share your location.

   If you don't want to share your location, tap Cancel.

Stop receiving new location sharing requests

Tap Me, then turn off Allow Friend Requests.

Add or remove a friend in Find My on iPad

In the Find My app , once you share your location with a friend, you can ask to see their location on a map.

Ask to see a friend's location

1. Tap People, then tap the name of the person whose location you want to see.

   If you don't see a person's name, make sure you're sharing your location with them.

2. Tap Ask To Follow Location.

After your friends receive and accept your request, you can see their locations.

Remove a friend

When you remove a friend, that person is removed from your People list and you are removed from theirs.

1. Tap People, then tap the name of the person you want to remove.

2. Tap Remove [*name*], then tap Remove.

Add a device to Find My on iPad

Before you can use the Find My app to locate a lost iPhone, iPad, iPod touch, Apple Watch, or AirPods, you need to connect the device to your Apple ID.

For your iPhone, iPad, iPod touch, Mac (with macOS 10.15 or later), and Apple Watch, Find My also includes a feature called Activation Lock that prevents anyone else from activating and using your device, even if it's completely erased.

Add an iPhone, iPad, or iPod touch

For a device with iOS 13, iPadOS 13, or later, follow the instructions below. For a device with iOS 12 or earlier.

1. On your iPhone, iPad, or iPod touch, go to Settings > [*your name*] > Find My.

If you're asked to sign in, enter your Apple ID. If you don't have one, tap "Don't have an Apple ID or forgot it?" then follow the instructions.

2. Tap Find My [*device*], then turn on Find My [*device*].

3. Turn on any of the following:

- *Find My network or Enable Offline Finding:* If your device is offline (not connected to Wi-Fi or cellular), Find My can locate it using the Find My network.

- *Send Last Location:* If your device's battery charge level becomes critically low, its location is sent to Apple automatically.

Add a Mac

1. On your Mac, choose Apple menu > System Preferences.

2. Do one of the following:

- *With macOS 10.15 or later:* Click Apple ID, then click iCloud.

- *With macOS 10.14 or earlier:* Click iCloud.

If you're asked to sign in, enter your Apple ID. If you don't have one, click "Don't have an Apple ID or forgot it?" then follow the instructions.

Select Find My Mac, then click Allow.

When you select Find My Mac on a Mac with macOS 10.15 or later, Offline Finding is turned on. Offline Finding allows your Mac to be located using Bluetooth when your Mac isn't connected to Wi-Fi. To turn off this option, click Options, then click Turn Off next to Offline Finding.

Add Apple Watch or AirPods

- *Apple Watch:* Pair your watch with an iPhone that you're signed in on with your Apple ID, or set up a watch for a family member. *AirPods:* Pair your AirPods with an iPhone, iPad, or iPod touch on which you're signed in with your Apple ID.

Add a family member's device

You can see your family members' devices in Find My if you set up Family Sharing first. Their devices appear below yours in the Devices list. You can't add friends' devices to Find My. Friends who lose a device can go to icloud.com/find and sign in with their Apple ID.

Locate a device in Find My on iPad

Use the Find My app to locate and play a sound on a missing iPhone, iPad, iPod touch, Mac, Apple Watch, or AirPods. In order to locate a device, you must turn on Find My [device] *before* it's lost.

Note: If you want to see how far away your devices are from you, make sure you turn on Precise Location for the Find My app.

If you lose your iPad and don't have access to the Find My app, you can locate or play a sound on your device using Find My iPhone on iCloud.com.

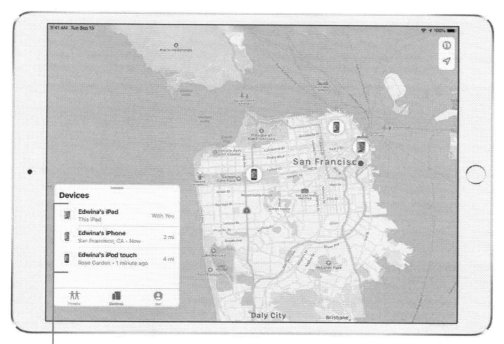

Tap a device to play a sound,
get directions, and more.

See the location of a device

Tap Devices, then tap the name of the device you want to locate.

- *If the device can be located:* It appears on the map so you can see where it is.

- *If the device can't be located:* You see "No location found" under the device's name. Under Notifications, turn on Notify When Found. You receive a notification once it's located.

Play a sound on your iPhone, iPad, iPod touch, Mac, or Apple Watch

1. Tap Devices, then tap the name of the device you want to play a sound on.

2. Tap Play Sound.

   - *If the device is online:* A sound starts after a short delay and gradually increases in volume, then plays for about two minutes. The device vibrates (if applicable). A Find My [*device*] alert appears on the device's screen.

     A confirmation email is also sent to your Apple ID email address.

   - *If the device is offline:* You see Sound Pending. The sound plays the next time the device connects to a Wi-Fi or cellular network.

Ask Siri. Say something like: "Help me find my iPad" or "Play a sound on my iPod touch."

Play a sound on your AirPods

1. Tap Devices, then tap the name of the AirPods you want to play a sound on.

2. Tap Play Sound. If your AirPods are separated, you can mute one by tapping Left or Right to find them one at a time.

- *If your AirPods are online:* They play a sound immediately (for two minutes).

    A confirmation email is also sent to your Apple ID email address.

- *If your AirPods are offline:* You receive a notification the next time your AirPods are our of their case and in range of your iPhone, iPad, or iPod touch.

Note: If your AirPods are in their case, they won't play a sound.

Stop playing a sound on a device

If you find your device and want to turn off the sound before it stops automatically, do one of the following:

- *iPhone, iPad, or iPod touch:* Press the power button or a volume button, or flip the Ring/Silent switch. If the device is locked, you can also unlock it, or swipe to dismiss the Find My [*device*] alert. If the device is unlocked, you can also tap OK in the Find My [*device*] alert.

- *Apple Watch:* Tap Dismiss in the Find My Watch alert, or press the Digital Crown or side button.

- *Mac:* Click OK in the Find My Mac alert.

- *AirPods:* Put your AirPods in their case and close the lid, or tap Stop in Find My.

Get directions to a device

You can get directions to a device's current location in the Maps app .

1. Tap Devices, then tap the name of the device you want to get directions to.

2. Tap Directions to open Maps.

3. Tap the route to get directions from your current location to the device's location.

Locate or play a sound on a friend's device

If your friend loses a device, they can locate it or play a sound on it by going to icloud.com/find and signing in with their Apple ID and password. If you set up Family Sharing, you can use Find My to locate a family member's missing device.

Mark a device as lost in Find My on iPad

Use the Find My app   to mark a missing iPhone, iPad, iPod touch, Apple Watch, or Mac as lost so that others can't access your personal information. In order to mark a device as lost, you must turn on Find My [device] *before* it's lost.

Add a custom message with your phone number.

What happens when you mark a device as lost?

- A confirmation email is sent to your Apple ID email address.

- You can display a custom message on the device's Lock Screen. For example, you may want to indicate that the device is lost or how to contact you.

- Your device doesn't display alerts or make noise when you receive messages or notifications, or if any alarms go off. Your device can still receive phone calls and FaceTime calls.

- Apple Pay is disabled for your device. Any credit or debit cards set up for Apple Pay, student ID cards, and Express Transit cards are removed from your device. Credit, debit, and student ID cards are removed even if your device is offline. Express Transit cards are removed the next time your device goes online.

- For an iPhone, iPad, iPod touch, or Apple Watch, you see your device's current location on the map as well as any changes in its location.

Mark a device as lost

If your device is lost or stolen, you can turn on Lost Mode for your iPhone, iPad, iPod touch, or Apple Watch, or lock your Mac.

1. Tap Devices, then tap the name of the lost device.

2. Under Mark As Lost, tap Activate.

3. Follow the onscreen instructions, keeping the following in mind:

   - *Passcode:* If your iPhone, iPad, iPod touch, or Apple Watch doesn't have a passcode, you're asked to create one now. For a Mac, you must create a numerical passcode, even if you already have a password set up on your Mac. This passcode is distinct from your password and is only used when you mark your device as lost.

   - *Contact Information:* If you're asked to enter a phone number, enter a number where you can be reached. If you're asked to enter a message, you may want to indicate that the device is lost or how to contact you. The number and message appear on the device's Lock Screen.

Tap Activate (for an iPhone, iPad, iPod touch, or Apple Watch) or Lock (for a Mac).

When the device has been marked as lost, you see Activated under the Mark As Lost section. If the device isn't connected to a Wi-Fi or cellular network when you mark it as lost, you see Pending until the device goes online again.

Change the Lost Mode message or email notifications for a lost device

After you mark your iPhone, iPad, iPod touch, or Apple Watch as lost, you can update your contact information or email notification settings.

1. Tap Devices, then tap the name of the lost device.

2. Under Mark As Lost, tap Pending or Activated.

3. Do any of the following:

   • *Change Lost Mode message:* Make any changes to the phone number or message.

   • *Get email updates:* Turn on Receive Email Updates if it's not already on.

   Tap Done.

Turn off Lost Mode for an iPhone, iPad, iPod touch, or Apple Watch

When you find your lost device, do either of the following to turn off Lost Mode:

- Enter your passcode on the device.

- In Find My, tap the name of the device, tap Pending or Activated under Mark as Lost, tap Turn Off Mark As Lost, then tap Turn Off.

Unlock a Mac

When you find your lost Mac, enter the numeric passcode on the device to unlock it (the one you set up when you marked your Mac as lost).

If you forget your passcode, you can recover it using Find My iPhone on iCloud.com.

Intro to Home on iPad

The Home app provides a secure way to control and automate HomeKit-enabled accessories, such as lights, locks, smart TVs, thermostats, window shades, smart plugs, and cameras. You can also view and capture video from supported security cameras, receive a notification when a supported doorbell camera recognizes someone at your door, and group multiple speakers to play the same audio. With Home, you can control any Works with Apple HomeKit accessory using iPad.

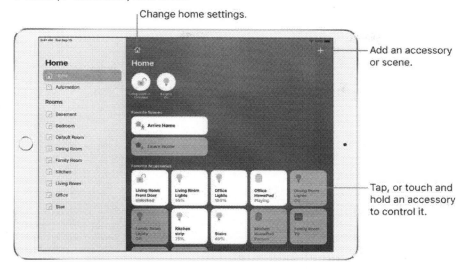

Change home settings.

Add an accessory
or scene.

Tap, or touch and
hold an accessory
to control it.

After you set up your home and its rooms, you can control accessories individually, or use scenes to control multiple accessories with one command. For example, you might create a scene called "wake up" that turns on lights in the kitchen, raises the thermostat, plays your morning playlist on the kitchen HomePod, and unlocks the front door.

To control your home automatically and remotely, you must have Apple TV (4th generation or later), HomePod, or iPad (with iOS 10.3, iPadOS 13, or later) that you leave at home. You can schedule scenes to run automatically at certain times, or when you activate a particular accessory (for example, when you unlock the front door). This also lets you, and others you invite, securely control your home while you're away.

Set up accessories with Home on iPad

The first time you open the Home app ⌂, the setup assistant helps you create a home, where you can add accessories and define rooms. If you already created a home using another HomeKit-enabled app, you'll skip this step.

Add an accessory to Home

Before you add an accessory such as a light or camera, be sure that it's connected to a power source, is turned on, and is using your Wi-Fi network.

1. Tap Home in the sidebar, then tap ⊕.

2. Tap Add Accessory, then follow the onscreen instructions.

   When you add an accessory, it's assigned to a default room, or a room you choose.

You may need to scan a QR code or enter an 8-digit HomeKit setup code found on the accessory itself (or its box or documentation). A supported smart TV displays a QR code for you to scan. You can assign the accessory to a room, and give it a name, and then use this name when controlling the accessory with Siri. You can also add suggested automations during set up.

When you set up Apple TV in tvOS and assign it to a room, it automatically appears in that room in the Home app on iPad.

Change an accessory's room assignment

1. In the sidebar, choose the room the accessory is currently assigned to.

   If it's not already assigned, look in Default Room.

2. Touch and hold the accessory's button, then swipe up or tap ⚙.

3. Tap Room, then choose a room.

Turn on Include in Favorites to add the accessory to the Home tab.

To rearrange your favorites, tap ⌂, tap Edit Screen, then drag the accessory buttons into the arrangement you want. Tap Done when you're finished.

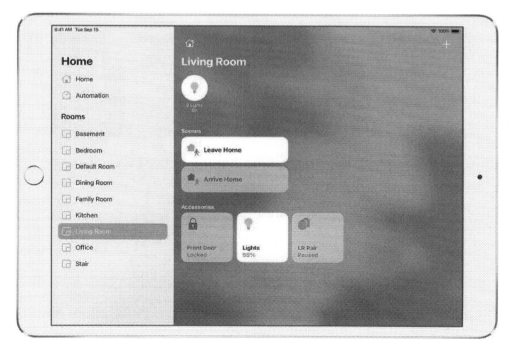

Organize rooms into zones

Group rooms together into a zone to easily control different areas of your home with Siri. For example, if you have a two-story home, you can assign the rooms on the first floor to a downstairs zone. Then you can say something to Siri like "Turn off the lights downstairs."

1. Tap , then tap Home Settings.

2. Tap Zone, then tap an existing zone, or tap Create New to add the room to a new zone.

Edit a room

You can change a room's name and wallpaper, add the room to a zone, or remove the room. When you remove the room, the accessories assigned to it move to Default Room.

1. Tap , then tap Room Settings.

2. Tap a room, then make your changes.

Get music, movies, TV shows, and more in the iTunes Store on iPad

Use the iTunes Store app to add music, movies, TV shows, and ringtones to iPad.

Note: You need an internet connection to use the iTunes Store. The availability of the iTunes Store and its features varies by country or region.

Find music, movies, and TV shows

1. In the iTunes Store, tap any of the following:

   - *Music, Movies, or TV Shows:* Browse by category. To refine your browsing, tap Genres at the top of the screen.

   - *Top Charts:* See what's popular on iTunes.

   - *Genius:* Browse recommendations based on what you bought from iTunes.

   - *Search:* Enter what you're looking for, then tap Search on the keyboard.

   Tap an item to see more information about it. You can preview songs, watch trailers for movies and TV shows, or tap ⬆️ to do any of the following:

   - *Share a link to the item:* Choose a sharing option.

   - *Give the item as a gift:* Tap Gift.

   - *Add the item to your wish list:* Tap Add to Wish List.

   To view your wish list, tap ☰, then tap Wish List.

Buy and download content

1. To buy an item, tap the price. If the item is free, tap Get.

   If you see ☁️⬇️ instead of a price, you already purchased the item, and you can download it again without a charge.

2. If required, authenticate your Apple ID with Face ID, Touch ID, or your passcode to complete the purchase.

3. To see the progress of a download, tap Downloads.

Get ringtones

1. Tap Music, tap Genres, scroll to the bottom, then tap Tones.

2. Browse by category or tap Top Charts to see what's popular.

3. Tap a ringtone to see more information or play a preview.

4. To buy a ringtone, tap the price.

Redeem or send an App Store & iTunes Gift Card

1. Tap Music, then scroll to the bottom.

2. Tap Redeem or Send Gift.

Get ringtones, text tones, and alert tones in the iTunes Store on iPad

In the iTunes Store app ⭐, you can purchase ringtones, text tones, and other alert tones for clock alarms and more.

Buy new tones

1. In the iTunes Store, tap Genres, then tap Tones.

2. Browse by category or tap Search to find a specific song or artist.

3. Tap a tone to see more information or play a preview.

4. To buy a tone, tap the price.

Redownload tones purchased with your Apple ID

If you bought tones on another device, you can download them again.

1. Go to Settings 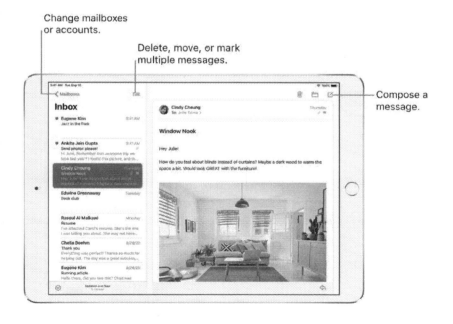 > Sounds.

2. Below Sounds and Vibration Patterns, tap any sound.

3. Tap Download All Purchased Tones. You might not see this option if you already downloaded all the tones that you purchased or if you haven't purchased any tones.

Write an email in Mail on iPad

With the Mail app , you can write and edit emails, and send and receive photos, videos, drawings, documents, and more.

Change mailboxes or accounts.

Delete, move, or mark multiple messages.

Compose a message.

Create an email message

Ask   Siri. Say something   like: "New   email   to   John Bishop" or "Email Simon and say I got the forms, thanks." Learn how to ask Siri.

Or do the following:

1. Tap ☑ .

2. Tap in the email, then type your message.

   With the onscreen keyboard, you can tap individual keys. Or pinch closed to use the smaller QuickType keyboard, then slide your finger from one letter to the next without lifting your finger. To change the formatting, tap Aa.

   You can change the font style, change the color of text, use a bold or italic stye, add a bulleted or numbered list, and more.

Reply to an email

1. Tap in the email, tap ↰, then tap Reply.

2. Type your response.

   With the onscreen keyboard, you can tap individual keys. Or pinch closed to use the smaller QuickType keyboard, then slide your finger from one letter to the next without lifting your finger.

Quote some text when you reply to an email

When you reply to an email, you can include text from the sender to clarify what you're responding to.

1. In the sender's email, touch and hold the first word of the text, then drag to the last wordTap ↰, then tap Reply, then type your message.

To turn off the indentation of quoted text, go to Settings > Mail > Increase Quote Level.

View maps on iPad

In the Maps app 🗺, you can find your location on a map and zoom in and out to see the detail you need.

To find your location, iPad must be connected to the internet, and Location Services must be on. On Wi-Fi + Cellular models, cellular data rates may apply.

Show your current location

Tap ◁.

Your position is marked in the middle of the map. The top of the map is north. To show your heading instead of north at the top, tap ◀. To resume showing north, tap ⋀ or 🧭.

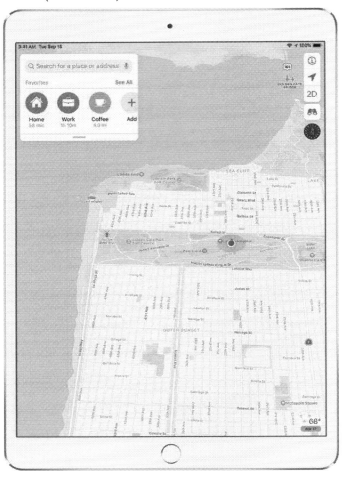

Choose between road, transit, and satellite views

Tap ⓘ, choose Map, Transit, or Satellite, then tap ✕.

If transit information is unavailable, tap View Routing Apps to use an app for public or other modes of transportation.

Move, zoom, and rotate a map

- *Move around in a map:* Drag the map.

- *Zoom in or out:* Double-tap and hold your finger to the screen, then drag up to zoom in or drag down to zoom out. Or, pinch open or closed on the map.

  The scale appears in the upper left while you're zooming. To change the unit of distance, go to Settings > Maps, then select In Miles or In Kilometers.

- *Rotate the map:* Touch and hold the map with two fingers, then rotate your fingers.

  To show north at the top of the screen after you rotate the map, tap .

View a 3D map

- *On a 2D road or transit map:* Drag two fingers up.

- *On a 2D satellite map:* Tap 3D near the upper right.

While viewing a 3D map, you can do the following:

- *Adjust the angle:* Drag two fingers up or down.

- *See buildings and other small features in 3D:* Zoom in.

- *Return to a 2D map:* Tap 2D near the upper right.

Allow Maps to use your precise location

To figure out where you are and provide accurate directions to your destinations, Maps works best when Precise Location is turned on.

To turn on Precise Location, do the following:

1. Go to Settings ⚙ > Privacy > Location Services.

2. Tap Maps, then turn on Precise Location.

Measure dimensions with iPad

On supported models, use the Measure app 📏 and your iPad camera to measure nearby objects and surfaces—you can manually set the start and end points of a measurement, have iPad automatically detect the dimensions of rectangular objects, and more.

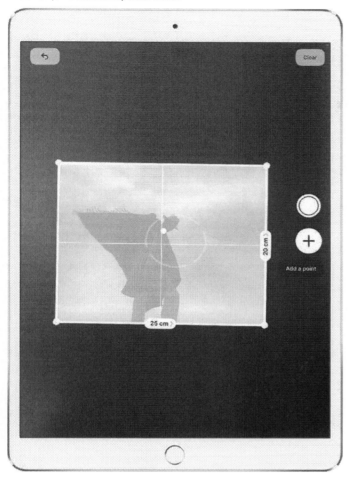

For best results, use Measure on well-defined objects located 0.5 to 3 meters (2 to 10 feet) from iPad.

Note: Measurements are approximate.

Start a measurement

1. Open Measure 📱, then use the iPad camera to slowly scan nearby objects.

2. Position iPad so that the object you want to measure appears on the screen.

Note: For your privacy, when you use Measure to take measurements, a green dot appears at the top of the screen to indicate your camera is in use.

Take an automatic rectangle measurement

1. When iPad detects the edges of a rectangular object, a box frames the object; tap the box to see the dimensions.

2. To take a photo of your measurement, tap ◯.

Take a manual measurement

1. Align the dot at the center of the screen with the point where you want to start measuring, then tap ⊕.

2. Slowly pan iPad to the end point, then tap ⊕ to see the measured length.

3. To take a photo of your measurement, tap ◯.

4. Take another measurement, or tap Clear to start over.

Use edge guides

On iPad Pro 11-inch (2nd generation) and iPad Pro 12.9-inch (4th generation), you can easily and precisely measure the heights and straight edges of furniture, countertops, and other objects using guide lines that appear automatically.

1. Position the dot at the center of the screen along the straight edge of an object until a guide appears.

2. Tap ⊕ where you want to begin measuring.

173

3. Slowly pan along the guide, then tap ⊕ at the endpoint to see the measured length.

4. To take a photo of your measurement, tap ◯.

Use Ruler view

On iPad Pro 11-inch (2nd generation) and iPad Pro 12.9-inch (4th generation), you can see more detail in your measurements with ruler view.

1. After measuring the distance between two points, move iPad closer to the measurement line until it transforms into a ruler, showing incremental inches and feet.

2. To take a0photo of your measurement, tap ◯.

Set up Messages on iPad

In the Messages app ◻, you can send text messages as SMS/MMS messages through your cellular service, or with iMessage over Wi-Fi or cellular service to people who use iPhone, iPad, iPod touch, or a Mac. Texts you send and receive using iMessage don't count against your SMS/MMS allowances in your cellular messaging plan, but cellular data rates may apply.

iMessage texts can include photos, videos, and other info. You can see when other people are typing, and send read receipts to let them know when you've read their messages. For security, messages sent using iMessage are encrypted before they're sent.

iMessage texts appear in blue bubbles, and SMS/MMS texts appear in green bubbles. See the Apple Support article About iMessage and SMS/MMS.

Sign in to iMessage

1. Go to Settings ⚙ > Messages.

2. Turn on iMessage.

Sign in to iMessage on your Mac and other Apple devices using the same Apple ID

If you sign in to iMessage with the same Apple ID on all your devices, all the messages that you send and receive on iPad also appear on your other Apple devices. Send a message from whichever device is closest to you, or use Handoff to start a conversation on one device and continue it on another.

1. On iPhone, iPad, or iPod touch, go to Settings ⚙ > Messages, then turn on iMessage.

2. On your Mac, open Messages, then do one of the following:

    • If you're signing in for the first time, enter your Apple ID and password, then click Sign In.

- If you signed in before and want to use a different Apple ID, choose Messages > Preferences, click iMessage, then click Sign Out.

With Continuity, you can send and receive SMS/MMS messages on iPad using the cellular connection on your iPhone.

Use Messages in iCloud

Go to Settings > [*your name*] > iCloud, then turn on Messages (if it's not already turned on).

Every message you send and receive on your iPad is saved in iCloud. And, when you sign in with the same Apple ID on a new device that also has Messages in iCloud turned on, all your conversations show up there automatically.

Because your messages and any attachments are stored in iCloud, you may have more free space on your iPad when you need it. Message bubbles, whole conversations, and attachments you delete from iPad are also deleted from your other Apple devices (iOS 11.4, iPadOS 13, macOS 10.13.5, or later) where Messages in iCloud is turned on.

Send and receive text messages on iPad

Use the Messages app to send and receive texts, photos, videos, and audio messages. You can also personalize your messages with animated effects, Memoji stickers, iMessage apps, and more.

Start a new conversation.

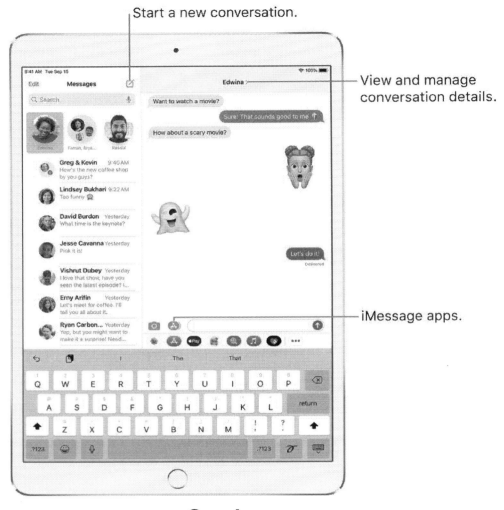

View and manage conversation details.

iMessage apps.

Send a message

You can send a text message to one or more people.

1. Tap 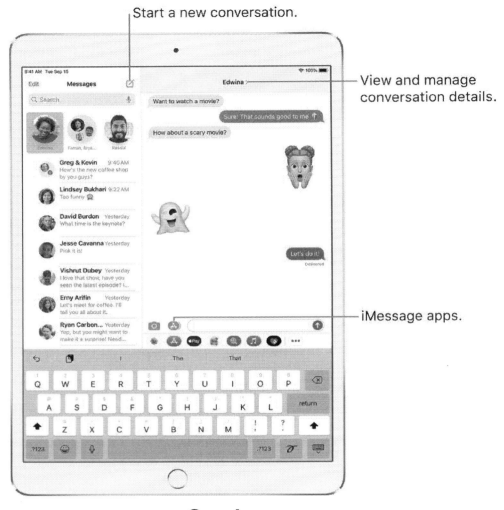 at the top of the screen to start a new message, or tap an existing message.

2. Enter the phone number, contact name, or Apple ID of each recipient. Or, tap ⊕, then choose contacts.

3. Tap the text field, type your message, then tap 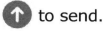 to send.

- A blue send button indicates the message will be sent with iMessage; a green send button indicates the message will be sent with SMS/MMS, or your cellular service.

- An alert ⊘ appears if a message can't be sent. Tap the alert to try sending the message again.

Tip: To see what time a message was sent or received, drag the message bubble to the left.

To view conversation details, tap the name or phone number at the top of the screen, then tap ⓘ. You can tap the contact to edit the contact card, share your location, view attachments, leave a group conversation, and more.

Reply to a message

Ask Siri. Say something like:

- "Send a message to Eliza saying how about tomorrow"

- "Read my last message from Bob"

- "Reply that's great news"

Tip: When you wear AirPods Pro, AirPods (2nd generation), or other supported headphones, Siri can read your incoming messages, and you can speak a reply for Siri to send (iPadOS 13.2 or later).

Or do the following:

1. In the Messages list, tap the conversation that you want to reply to.

   To search for contacts and content in conversations, pull down the Messages list and enter what you're looking for in the search field. Or, choose from the suggested contacts, links, photos, and more.

2. Tap the text field, then type your message.

   Tip: To replace text with emoji, tap 😃 or 🌐 , then tap each highlighted word.

3. Tap ⬆ to send your message.

You can quickly reply to a messages with a Tapback expression (for example, a thumbs up or a heart). Double-tap the message bubble that you want to respond to, then select a Tapback.

Pin a conversation

You can pin specific conversations to the top of the Messages list so the people you contact most always come first in the list.

Do any of the following:

- Swipe right on a conversation, then tap 📌 .

- Touch and hold a conversation, then drag it to the top of the list.

Unpin a conversation

You can unpin specific conversations at the top of the Messages list.

Do any of the following:

- Touch and hold a conversation, then drag the message to the bottom of the list.

- Touch and hold a conversation, then tap 📌.

Share your name and photo

In Messages, you can share your name and photo when you start or respond to a new message. Your photo can be a Memoji or custom image. When you open Messages for the first time, follow the instructions on your iPad to choose your name and photo.

To change your name, photo, or sharing options, open Messages, tap ⋯, tap Edit Name and Photo, then do any of the following:

- *Change your profile image*: Tap Edit, then choose an option.

- *Change your name*: Tap the text fields where your name appears.

- *Turn sharing on or off*: Tap the button next to Name and Photo Sharing (green indicates that it's on).

- *Change who can see your profile*: Tap an option below Share Automatically (Name and Photo Sharing must be turned on).

Your Messages name and photo can also be used for your Apple ID and My Card in Contacts.

Switch from a Messages conversation to a FaceTime or audio call

In a Messages conversation, you can initiate a FaceTime or audio call with the person you're chatting with in Messages.

1. In a Messages conversation, tap the profile picture or the name at the top of the conversation.

2. Tap FaceTime or audio.

Send and receive text messages on iPad

Use the Messages app to send and receive texts, photos, videos, and audio messages. You can also personalize your messages with animated effects, Memoji stickers, iMessage apps, and more.

Start a new conversation.

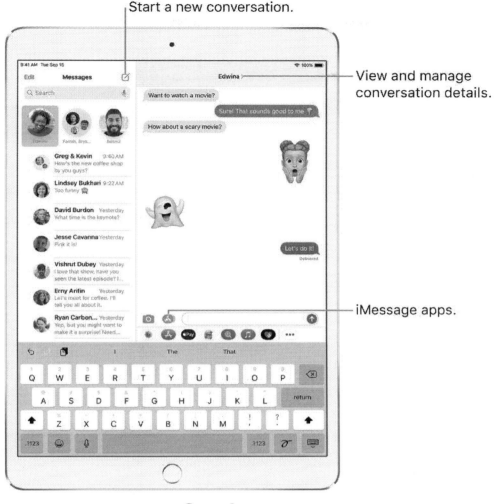

View and manage
conversation details.

iMessage apps.

Send a message

You can send a text message to one or more people.

1. Tap 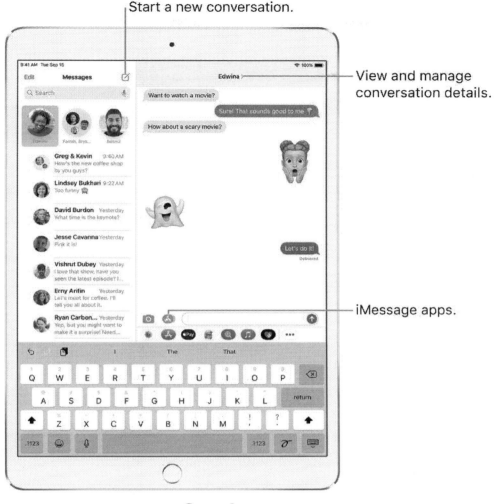 at the top of the screen to start a new message, or tap an existing message.

2. Enter the phone number, contact name, or Apple ID of each recipient. Or, tap ⊕, then choose contacts.

3. Tap the text field, type your message, then tap ⬆ to send.

- A blue send button indicates the message will be sent with iMessage; a green send button indicates the message will be sent with SMS/MMS, or your cellular service.

- An alert ⚠ appears if a message can't be sent. Tap the alert to try sending the message again.

Tip: To see what time a message was sent or received, drag the message bubble to the left.

To view conversation details, tap the name or phone number at the top of the screen, then tap ℹ. You can tap the contact to edit the contact card, share your location, view attachments, leave a group conversation, and more.

Reply to a message

Ask Siri. Say something like:

- "Send a message to Eliza saying how about tomorrow"

- "Read my last message from Bob"

- "Reply that's great news"

Tip: When you wear AirPods Pro, AirPods (2nd generation), or other supported headphones, Siri can read your incoming messages, and you can speak a reply for Siri to send (iPadOS 13.2 or later).

Or do the following:

1. In the Messages list, tap the conversation that you want to reply to.

   To search for contacts and content in conversations, pull down the Messages list and enter what you're looking for in the search field. Or, choose from the suggested contacts, links, photos, and more.

2. Tap the text field, then type your message.

   Tip: To replace text with emoji, tap 😃 or 🌐 , then tap each highlighted word.

3. Tap ⬆ to send your message.

You can quickly reply to a messages with a Tapback expression (for example, a thumbs up or a heart). Double-tap the message bubble that you want to respond to, then select a Tapback.

Pin a conversation

You can pin specific conversations to the top of the Messages list so the people you contact most always come first in the list.

Do any of the following:

- Swipe right on a conversation, then tap 📌 .

- Touch and hold a conversation, then drag it to the top of the list.

Unpin a conversation

You can unpin specific conversations at the top of the Messages list.

Do any of the following:

- Touch and hold a conversation, then drag the message to the bottom of the list.

- Touch and hold a conversation, then tap ✕.

Share your name and photo

In Messages, you can share your name and photo when you start or respond to a new message. Your photo can be a Memoji or custom image. When you open Messages for the first time, follow the instructions on your iPad to choose your name and photo.

To change your name,   photo, or sharing options, open Messages, tap ●●●, tap Edit Name and Photo, then do any of the following:

- *Change your profile image*: Tap Edit, then choose an option.

- *Change your name*: Tap the text fields where your name appears.

- *Turn sharing on or off*: Tap the button next to Name and Photo Sharing (green indicates that it's on).

- *Change who can see your profile*: Tap an option below Share Automatically (Name and Photo Sharing must be turned on).

Your Messages name and photo can also be used for your Apple ID and My Card in Contacts.

Switch from a Messages conversation to a FaceTime or audio call

In a Messages conversation, you can initiate a FaceTime or audio call with the person you're chatting with in Messages.

1. In a Messages conversation, tap the profile picture or the name at the top of the conversation.

2. Tap FaceTime or audio.

View albums, playlists, and more in Music on iPad

In the Music app 🎵, the library includes music you added or downloaded from Apple Music, music and videos you synced to iPad, TV shows and movies you added from Apple Music, and your iTunes Store purchases.

Choose a sorting method.

Browse and play your music

1. Under the Library heading in the sidebar, tap a category, such as Albums or Songs; tap Downloaded to view only music stored on iPad.

2. Type in the search field to filter your results and find what you're looking for.

3. Tap an item, then tap Play, or tap Shuffle to shuffle an album or playlist.

   You can also touch and hold the album art, then tap Play.

To change the list of categories, tap Edit in the sidebar, then select categories you'd like to add, such as Genres and Compilations. Tap any existing categories to remove them.

Sort your music

1. Under the Library heading in the sidebar, tap Playlists, Albums, Songs, TV & Movies, or Music Videos.

2. Tap Sort, then choose a sorting method, such as title, artist, recently added, or recently played.

Play music shared on a nearby computer

If a computer on your network shares music through Home Sharing, you can stream its music to your iPad.

1. Go to Settings > Music, tap Sign In below Home Sharing, then sign in with your Apple ID.

2. Open the Music app , tap Edit in the sidebar, select Home Sharing, then tap Done.

3. Tap Home Sharing, then choose a shared library.

Remove Apple Music songs from iPad

Go to Settings > Music, then turn off Sync Library.

The songs are removed from iPad, but remain in iCloud. Music you purchased or synced also remains.

Play music on iPad

Use Now Playing in the Music app to show lyrics and play, pause, skip, shuffle, and repeat songs. You can also use Now

Playing to view album art and choose what plays next in the queue.

Control playback

Tap the player at the bottom right to show the Now Playing screen, where you can use these controls:

Control   Description

▶   Play the current song.

❚❚   Pause playback.

▶▶   Skip to the next song. Touch and hold to fast-forward through the current song.

◀◀   Return to the song's beginning. Tap again to play the previous song in an album or playlist. Touch and hold to rewind through the current song.

🔁   Tap to repeat an album or playlist. Double-tap to repeat a single song.

Tap to play your songs in random order. Tap again to turn off shuffle.

— Hide the Now Playing Screen button.

••• Tap for more options.

 Show time-synced lyrics (lyrics not available for all songs).

 Stream music to Bluetooth or AirPlay-enabled devices.

 See the queue.

Tap to hide Now Playing.

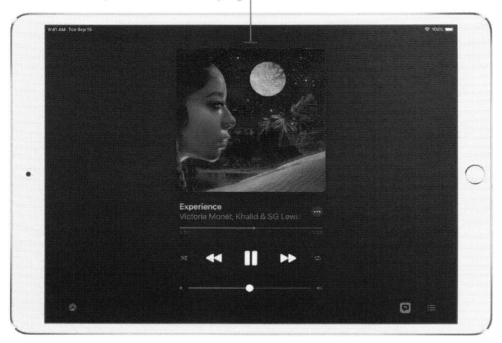

Adjust the volume, see song details, and more

The Now Playing screen contains additional options for controlling and accessing music.

- *Adjust volume:* Drag the volume slider.

  You can also use the volume buttons on the side of iPad.

- *Navigate to the artist, album, or playlist:* Tap the artist name below the song title, then choose to go to the artist, album, or playlist.

- *Scrub to any point in a song:* Drag the playhead.

See time-synced lyrics

Time-synced lyrics appear for many songs in Apple Music.

Tap the player to open Now Playing. Lyrics scroll in time with the music.

To hide lyrics, tap 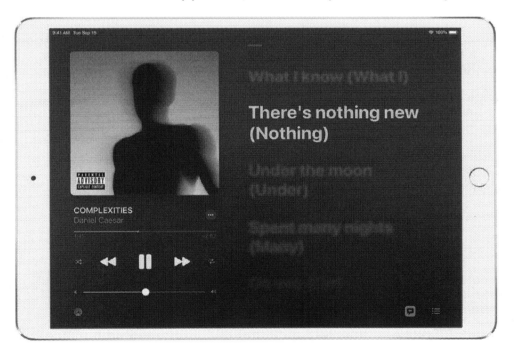.

Tip: Tap a specific lyric—the first line of the chorus, for example—to go to that part of the song.

To see all of a song's lyrics, tap ●●● , then tap View Full Lyrics.

Note: You need an Apple Music subscription to view lyrics.

Get audio controls from the Lock screen or when using another app

Open Control Center, then tap the audio card.

Stream music to Bluetooth or AirPlay-enabled devices

1. Tap the player to open Now Playing.

2. Tap , then choose a device.

Note: You can play the same music on multiple AirPlay 2-enabled devices, such as two or more HomePod speakers. You can also pair two sets of AirPods to one iPad and enjoy the same song or movie along with a friend.

Get started with News on iPad

The News app collects all the stories you want to read, from your favorite sources, about the topics that interest you most. To personalize News, you can choose from a selection of publications (called *channels*) and topics such as Entertainment, Food, and Science.

The more you read, the better News understands your interests. Siri learns what's important to you and suggests stories you might like. You can also receive notifications about important stories from channels you follow.

You can also subscribe to Apple News+, which includes hundreds of magazines, popular newspapers, and premium digital publishers.

The News app and Apple News+ aren't available in all countries, regions, or languages.

Note: You need0a Wi-Fi or cellular connection to use News.

Personalize your news

When you follow a channel or topic, related stories appear more often in the Today feed, and the channel or topic appears below Following in the sidebar.

1. Open News for the first time, then, in the sidebar, tap ⊕ for channels or topics you want to follow.

   If there are channels or topics that you don't want to appear in your feed, tap ⋯ to block them or to stop News from suggesting them.

   When you block a channel or topic, its stories are filtered out of the Today feed and Today widget. However, you may see stories from channels you've blocked in Top Stories and other locations that feature stories chosen by Apple News editors. Tap Following > Blocked Channels & Topics in the sidebar to see the channels and topics you've blocked.

2. Tap Discover Channels & Topics at the bottom of the sidebar, then tap ⊕ for each channel you want to follow.

To stop following a channel or topic, swipe it to the left, then tap Unfollow.

To easily follow specific channels and topics, tap the search field at the top of the sidebar, enter the channel or topic's name, then tap ⊕ in the results below.

Get notifications

Some channels you follow may send notifications about important stories.

1. At the bottom of the sidebar, tap Notifications & Email below Manage.

2. Turn on notifications for your preferred channels.

If you subscribe to Apple News+, you can get notifications when new issues are available.

Receive Apple News newsletters

You can choose to receive newsletters in your email inbox from Apple News editors highlighting top stories. Newsletters can be personalized based on your interests.

1. Swipe to the bottom of the sidebar, then tap Notifications & Email.

2. Swipe to the bottom of the window, then turn on Apple News Newsletter.

To stop receiving newsletters, return to the Notifications & Email window, then turn off Apple News Newsletter.

Note: Personalized newsletters aren't available in all countries or regions.

See stories only from the channels you follow

Go to Settings ⚙, tap News, turn on Restrict Stories in Today, then confirm your choice.

Note: Restricting stories significantly limits the variety of stories that appear in the Today feed and all other feeds. For example, if you restrict stories and follow only one entertainment-related channel, your Entertainment topic feed will contain stories only from that channel. When you restrict stories, you don't see Top Stories and Trending Stories.

Take notes on iPad

Use the Notes app 📝 to jot down quick thoughts or organize detailed information with checklists, images, web links, scanned documents, handwritten notes, and sketches.

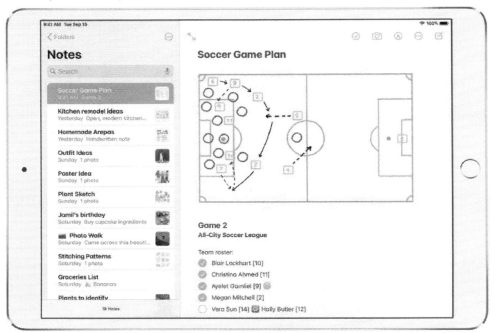

Create and format a new note

Ask Siri. Say something like: "Create a new note

Or do the following:

1. Tap , then enter your text.

   The first line of the note becomes the note's title.

2. To change the formatting, tap Aa.

   You can apply a heading style, bold or italic font, a bulleted or numbered list, and more.

3. To save the note, tap Done.

Tip: To choose a default style for the first line in all new notes, go to Settings ⚙ > Notes > New Notes Start With.

You can quickly create a note or resume work on your last note by tapping the Lock Screen with Apple Pencil (on supported models). On other iPad models, you can do this if you add Notes to Control Center. To change the Lock Screen behavior, go to Settings ⚙ > Notes > Access Note from Lock Screen.

Add a checklist

In a note, tap ✓, then do any of the following:

- *Add items to the list:* Enter text, then tap return to enter the next item.

- *Increase or decrease the indentation:* Swipe right or left on the item.

- *Mark an item as complete:* Tap the empty circle next to the item to add a checkmark.

- *Reorder an item:* Touch and hold the empty circle or checkmark next to the item, then drag the item to a new position in the list.

- *Manage items in the list:* Tap the list to see the menu, tap ▶, tap Checklist, then tap Check All, Uncheck All, Delete Checked, or Move Checked to Bottom.

To automatically sort checked items to the bottom in all your notes, go to Settings ⚙ > Notes > Sort Checked Items, then tap Automatically.

Add or edit a table

In a note, tap ⊞, then do any of the following:

- *Enter text:* Tap a cell, then enter your text. To start another line of text in the cell, touch and hold the Shift key and tap next.

- *Move to the next cell:* Tap next. When you reach the last cell, tap next to start a new row.

- *Format a row or column:* Tap a row or column selection handle, then choose a style, such as bold, italic, underline, or strikethrough.

- *Add or delete a row or column:* Tap a row or column selection handle, then choose to add or delete.

- *Move a row or column:* Touch and hold a row or column selection handle, then drag it to a new position.

- *See more columns:* If the table becomes wider than your screen, swipe right or left on the table to see all the columns.

To remove the table and convert its contents to text, tap a cell in the table, tap ⊞, then tap Convert to Text.

Change the Notes display on your iPad

- *Make the note fill the screen:* While viewing a note in landscape orientation, tap ⬉ or turn iPad to portrait orientation.

- *See your accounts, notes list, and selected note all at once:* (iPad Pro   12.9-inch)   Turn   iPad to   landscape orientation, then tap ▣ .

Draw or write in Notes on iPad

Use the Notes app ▭ to draw a sketch or jot a handwritten note with Apple Pencil (on supported models) or your finger. You can choose from a variety of Markup tools and colors and draw straight lines with the ruler.

Draw or write in a note

1. Start drawing or writing in the note with Apple Pencil. Or to draw or write with your finger, tap ⒶＡ .

2. Do any of the following:

   - *Change color or tools:* Use the Markup tools.

   - *Adjust the handwriting area:* Drag the resize handle (on the left) up or down.

- *Transcribe your handwriting to typed text as you write with Apple Pencil:* Tap the Handwriting tool ⌇ (to the left of the pen), then start writing.

To learn more about writing notes with Apple Pencil,

Tip: You can search handwritten text (in supported languages) in Notes. If the note doesn't have a title, the first line of handwritten text becomes the suggested title. To edit the title, scroll to the top of the note, then tap Edit.

Select and edit drawings and handwriting

With Smart Selection, you can select handwritten text using the same gestures you use for typed text. You can move, copy, or delete the selection within the note. You can even paste it as typed text in another note or app.

Note: Smart Selection and handwriting transcription work if the system language of your iPad is set to English, Simplified Chinese, or Traditional Chinese in Settings ⚙ > General > Language & Region > iPad Language.

1. In the Markup toolbar, tap the Lasso tool ⌇ (between the eraser and ruler).

2. Select drawings and handwriting with Apple Pencil or your finger using any of the following methods:

   - Touch and hold, then drag to expand the selection.

- Double-tap to select a word.

- Triple-tap to select a sentence.

Adjust the selection by dragging the handles if necessary.

Tap the selection, then choose Cut, Copy, Delete, Duplicate, Copy as Text, or Insert Space Above.

If you choose Copy as Text, you can paste the transcribed text in another note or another app.

Use handwritten addresses, phone numbers, dates, and more

A yellow underline appears below handwritten text that's recognized as a street or email address, phone number, date, or other data. Tap the underlined text to take action on it. For example, you can see a street address in Maps, compose an email, call a phone number, or add a calendar event.

Note: Data detection works if the system language of your iPad is set to English, Simplified Chinese, or Traditional Chinese in Settings > General > Language & Region.

Show lines or grids in notes

- *In an existing note:* Tap , tap Lines & Grids, then choose a style.

- *Choose the default style for all new notes:* Go to Settings > Notes > Lines & Grids.

ake photos in Photo Booth on iPad

Use the Photo Booth app 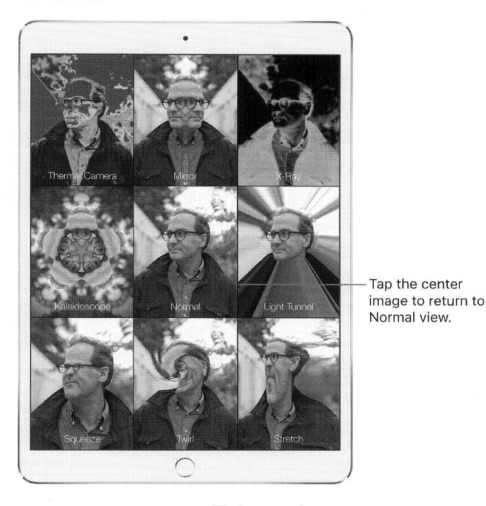 to take selfies and photos with fun effects.

Tap the center image to return to Normal view.

Take a photo

Photo Booth uses the front camera to display the subject in multiple tiles, each with a different effect, on the iPad screen. Effects include Kaleidoscope, Stretch, X-Ray, and more. The tile in the center of the screen displays Normal view.

1. Aim iPad at your subject to frame your shot.

2. Tap the tile of the effect you want to capture, then do any of the following:

   - *Switch between the front and rear-facing cameras:* Tap at the bottom of the screen.

   - *Change the effect:* Tap at the bottom left of the screen.

     With some effects, you can drag your finger across the screen, or pinch, swipe, or rotate the image to alter the appearance.

     Tap the shutter button to take the shot.

When you take a photo, iPad makes a shutter sound. You can use the volume buttons on the side of iPad to control the volume of the shutter sound.

Note: In some regions, sound effects are played even if the Side Switch (available on some models) is set to silent.

View photos and videos on iPad

Use the Photos app to view your photos and videos organized by years, months, days, and all photos in your photo library. Swipe from the left edge of the screen, or tap , to open the sidebar and find photos organized by different categories. You can also use Photos to create albums and then share them with friends and family.

Browse your library

The photos and videos on your iPad are organized in your photo library by Years, Months, Days, and All Photos. You can rediscover your best shots in Years, relive significant events in Months, focus on unique photos in Days, and view everything in All Photos.

To browse your photo library, tap the buttons at the top of the screen to choose a view:

- *Years:* Highlights your best shots from a specific year in your photo library.

- *Months:* Presents collections of photos that you took throughout a month, organized by significant events—like a family outing, birthday party, or trip.

- *Days:* Shows your best shots, grouped by the time or place the photos were taken.

- *All Photos:* Displays all of your photos and videos; tap ⬤
  to zoom in or out, view photos by aspect ratio or square, filter photos, or see photos on a map.

Photos removes duplicate photos and clutter (such as screenshots, whiteboards, and receipts) from Days, Months, and Years views. Every photo is shown in All Photos.

View individual photos

Tap a photo thumbnail to view it in full screen, then do any of the following:

- *Zoom in or out:* Double-tap or pinch out to zoom in—while zoomed in, drag to see other parts of the photo; double-tap or pinch closed to zoom out.

- *Share:* Tap ⬆️, then choose how you want to share. *Add to favorites:* Tap ♡ to add the photo to your Favorites album.

Swipe to browse
through your photos.

Tap a thumbnail to view a photo.

Tap ❮ or drag the photo down to continue browsing or return to the search results.

Add captions and view photo and video details

Captions add context to your photos and videos, and you can search for photos and videos by captions when you use Search. Select a photo or video, then swipe up to add a caption or view a caption in the text field below the image.

When you swipe up on a photo or video, you also see the following details:

- Effects you can add to a Live Photo.

- People identified in your photo.

- Where the photo was taken.

- A link to view other photos taken nearby.

In Shared Albums, you can add comments and likes to photos and videos. Your comments and likes are shared with the album subscribers.

Play a Live Photo

A Live Photo , which can be taken on supported models, is a moving image that captures the moments just before and after a picture is taken.

1. Open a Live Photo.

2. Touch and hold the photo to play it.

Tip: To see all of your Live Photos, swipe from the left edge of the screen or tap to show the sidebar, then tap Live Photos under Media Types.

View photos in a Burst shot

Burst mode in Camera takes multiple high-speed photos so that you have a range of photos to choose from. In Photos, Burst shots are saved together in a single photo thumbnail. You can view each photo in the Burst, then select your favorites to save separately.

1. Open a Burst photo.

2. Tap Select, then swipe through the collection of photos.

3. To save specific photos, tap each photo to select it, then tap Done.

4. Tap Keep Everything to keep the Burst and the photos you selected, or tap Keep Only [*number of*] Favorites to keep only the ones you selected.

Tip: To see all of your Burst shots, swipe from the left edge of the screen or tap [image] to show the sidebar, then tap Bursts under Media Types.

Play a video

As you browse your photo library, videos auto-play while you scroll. Tap a video to begin playing it in full screen without sound, then do any of the following:

- Tap the player controls below the video to pause, play, unmute, and mute; tap the screen to hide the player controls.

- Double-tap the screen to toggle between full screen and fit-to-screen.

Play and customize a slideshow

A slideshow is a collection of your photos, formatted and set to music.

1. View photos by All Photos or Days, then tap Select.

2. Tap each photo you want to include in the slideshow, then tap [image].

3. From the list of options, tap Slideshow.

4. Tap the screen, then tap Options to change the slideshow theme, music, and more.

Find podcasts on iPad

Use the Podcasts app to find and play free shows—similar to radio or TV shows—about science, news, politics, comedy, and more. If you subscribe to a show, iPad automatically downloads new episodes as they're released.

Ask Siri. Say something like: "Find Voyage To The Stars podcast."

Find and subscribe to shows

- *Discover shows:* Tap Browse to see Featured shows or Top Charts. You can also browse by categories or content providers.

- *Search by title, person, or topic:* Tap Search, then enter what you're looking for.

- *Subscribe to a show:* Tap the show to see its information page, then tap Subscribe.

Listen to your subscribed shows

Tap Listen Now, then select an episode from one of the following:

- *Up Next:* Pick up where you left off in a show, including resuming a previous episode.

- *Latest Episodes:* Find new episodes from shows you're already subscribed to.

- *Notifications:* To get notifications whenever a new episode is available, tap 🔔.

Play podcasts on iPad

In the Podcasts app 📻, you can play, pause, or skip ahead using the playback controls, set a sleep timer, and stream content to another device.

Ask Siri. Say something like: "Play the newest episode of 'The Daily' podcast.

Play a podcast

1. Tap an episode.

2. For more playback controls, tap the player to open the Now Playing screen.

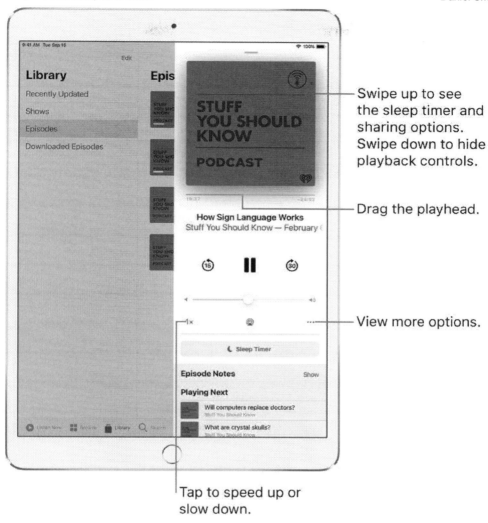

Swipe up to see the sleep timer and sharing options. Swipe down to hide playback controls.

Drag the playhead.

View more options.

Tap to speed up or slow down.

Use any of the following controls:

Control   Description

▶   Play

❚❚   Pause

Control   Description

Jump back 15 seconds

Jump forward 30 seconds

1×   Choose a faster or slower playback speed

Stream the audio to other devices

•••   Choose more actions such as sharing the
episode or adding it to your queue.

Tip: To jump to a specific time in the episode, drag the track
position slider below the podcast artwork.

Ask Siri. Say something like: "Skip ahead 3 minutes." Learn
how to ask Siri.

3. Swipe up on the Now Playing screen to see the sleep timer,
episode notes, and your Up Next queue.

Set reminders on iPad

In the Reminders app ⦚, you can easily create and organize
reminders to keep track of all of life's to-dos. Use it for shopping

lists, projects at work, tasks around the house, and anything else you want to track. Create subtasks, set flags, add attachments, and choose when and where to receive reminders. You can also use smart lists to automatically organize your reminders.

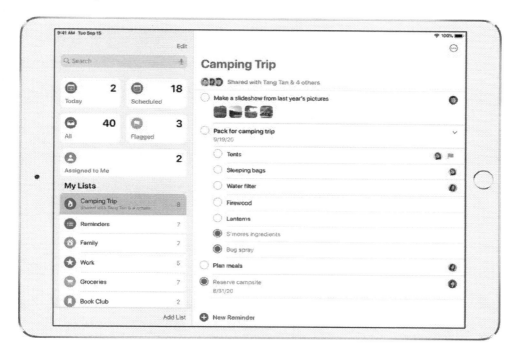

Keep your reminders up to date on all your devices with iCloud

Go to Settings 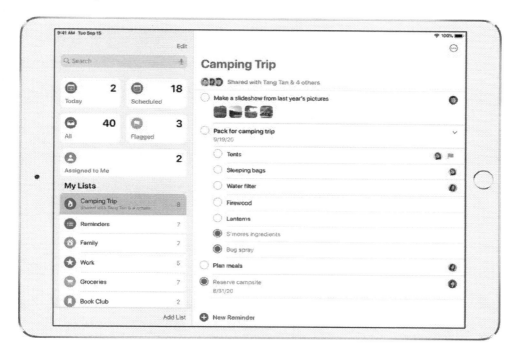 > [*your name*] > iCloud, then turn on Reminders.

Your iCloud reminders—and any changes you make to them— appear on your iPhone, iPad, iPod touch, Apple Watch, and Mac where you're signed in with your Apple ID.

Note: If you've been using Reminders on iOS 12 or earlier, you may need to upgrade your iCloud reminders to use features such as attachments, flags, subtasks, grouped lists, list colors and icons, and more. To upgrade, tap the Upgrade button next to your iCloud account in Reminders. (You may need to tap Lists at the top left to see your iCloud account.)

Upgraded reminders are not backward compatible with the Reminders app in earlier versions of iOS and macOS.

Add a reminder

Ask Siri. Say something like: "Add artichokes to my groceries list."

Or do the following in a list:

1. Tap New Reminder, then enter text. Or if you have Apple Pencil, write in the text field.

2. Use the quick toolbar above the keyboard to do any of the following:

   - *Schedule a date and time:* Tap 📅, then choose when you want to be reminded.

   - *Add a location:* Tap ⌁, then choose where you want to be reminded—for example, when you arrive home or get into a car with a Bluetooth connection to your iPad.

     Note: To receive location-based reminders, you must allow Reminders to use your precise location. Go to

Settings ⚙ > Privacy > Location Services, turn on Location Services, tap Reminders, choose While Using the App, then turn on Precise Location.

- *Assign the reminder:* (Available in shared lists) Tap ⃝ , then choose a person on the shared list (including yourself).

- *Set a flag:* Tap ⚑ to mark an important reminder.

- *Attach a photo or scanned document:* Tap 📷, then take a new photo, choose an existing photo from your photo library, or scan a document.

To add more details to the reminder, tap ⓘ, then do any of the following:

- *Add notes:* In the Notes field, enter more info about the reminder.

- *Add a web link:* In the URL field, enter a web address. Reminders displays the link as a thumbnail that you can tap to go to the website.

- *Get a reminder when chatting with someone in Messages:* Turn on "When messaging," then choose someone from your contacts list. The reminder appears the next time you chat with that person in Messages.

- *Set a priority:* Tap Priority, then choose an option.

    Tap Done.

Tip: With OS X 10.10 or later, you can hand off reminders you're editing between your Mac and iPad.

Mark a reminder as complete

Tap the empty circle next to a reminder to mark it as complete and hide it.

To unhide completed reminders, tap ⬤, then tap Show Completed.

Edit multiple reminders all at once

1. Tap ⬤, tap Select Reminders, then select the reminders you want to edit. Or drag two fingers over the reminders you want to edit.

2. Use the buttons at the bottom of the screen to complete, flag, add a date and time, move, assign, or delete the selected reminders.

Move or delete reminders

- *Reorder reminders in a list:* Touch and hold a reminder you want to move, then drag it to a new location.

- *Make a subtask:* Swipe right on the reminder, then tap Indent. Or drag a reminder onto another reminder.

   If you delete or move a parent task, the subtasks are also deleted or moved. If you complete a parent task, the subtasks are also completed.

- *Move a reminder to a different list:* Tap the reminder, tap ⓘ, tap List, then choose a list.

- *Delete a reminder:* Swipe left on the reminder, then tap Delete.

  Tip: To delete a reminder with Apple Pencil, just scribble over the reminder.

  To recover a deleted reminder, shake to undo or double-tap with three fingers.

Change your Reminders settings

1. Go to Settings ⚙ > Reminders.

2. Choose options such as the following:

   - *Accounts:* Add an account such as iCloud, Microsoft Exchange, or Yahoo.

   - *Default List:* Choose the list for new reminders you create outside of a specific list, such as reminders you create using Siri.

   - *Today Notification:* Set a time to show notifications for all-day reminders that have been assigned a date without a time.

   - *Show as Overdue:* The scheduled date turns red for overdue all-day reminders.

- *Mute Notifications:* Turn off notifications for assigned reminders.

Browse the web using Safari on iPad

With the Safari app , you can browse the web, add webpages to your reading list to read later, and add page icons to the Home Screen for quick access. If you sign in to iCloud with the same Apple ID on all your devices, you can see pages you have open on other devices, and keep your bookmarks, history, and reading list up to date on all your devices.

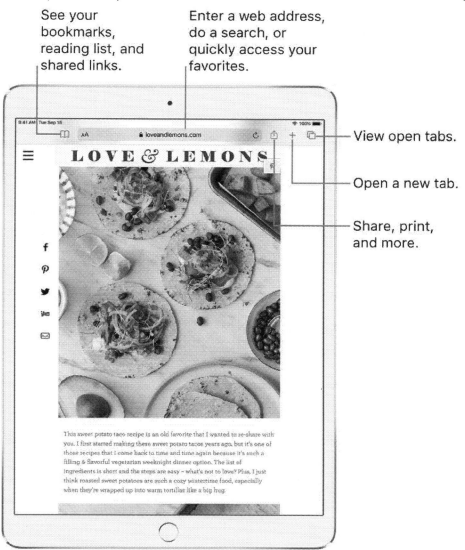

See your bookmarks, reading list, and shared links.

Enter a web address, do a search, or quickly access your favorites.

View open tabs.

Open a new tab.

Share, print, and more.

View websites with Safari

You can easily navigate a webpage with a few taps.

- *Get back to the top:* Double-tap the top edge of the screen to quickly return to the top of a long page.

- *See more of the page:* Turn iPad to landscape orientation.

- *Refresh the page:* Tap ↻ next to the address in the search field.

- *Share links:* Tap ⬆️

Change text size, display, and website settings

On iPad, Safari shows a website's desktop version that's automatically scaled for the iPad display and optimized for touch input.

Use the View menu to increase or decrease the text size, switch to Reader view, specify privacy restrictions, and more.

To open the View menu, tap AA on the left side of the search field, then do any of the following:

- *Change the font size:* Tap the large A to increase the font size or tap the small A to decrease it.

- *View the webpage without ads or navigation menus:* Tap Show Reader View (if available).

- *Hide the search field:* Tap Hide Toolbar (tap the top of the screen to get it back).

- *View the mobile version of the webpage:* Tap Request Mobile Website (if available).

- *Set display and privacy controls for each time you visit this website:* Tap Website Settings.

View two pages side-by-side in Split View

Use Split View to open two Safari pages side-by side.

- *Open a blank page in Split View:* Touch and hold ⬚, then tap Open New Window.

- *Open a link in Split View:* Touch and hold the link, then tap Open in New Window.

- *Move a window to the other side of Split View:* Touch and hold the top of the window, then drag left or right.

- *Close tabs in a Split View window:* Touch and hold ⬚.

- *Leave Split View:* Drag the divider over the window you want to close.

Preview website links

Touch and hold a link in Safari to see a preview of the link without opening the page. To open the link, tap the preview, or choose another option.

To close the preview and stay on the current page, tap anywhere outside the preview.

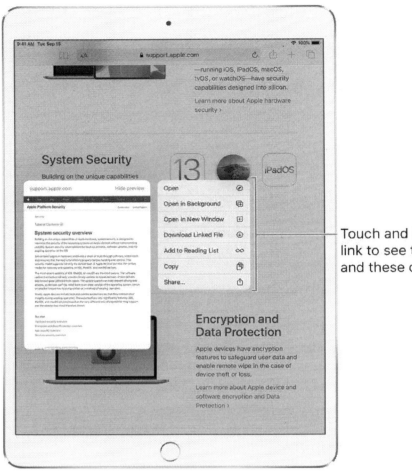

Touch and hold a link to see the URL and these options.

Translate a webpage

When you encounter a webpage that's in another language, you can use Safari to translate the text (beta; not available in all languages or regions).

When viewing a webpage in another language, tap $A\!A$, then tap .

Manage downloads

Tap ⊕ to check the status of a file you're downloading, to access downloaded files quickly, or to drag a downloaded file onto another file or into an email you're working on.

You can download files in the background while you continue to use Safari.

Use keyboard shortcuts

You can navigate in Safari using keyboard shortcuts on an external keyboard.

To view available keyboard shortcuts, press and hold the Command key.

Search for websites using Safari on iPad

In the Safari app 🧭, enter a URL or a search term to find websites or specific information.

Search the web

1. Enter a search term, phrase, or URL in the search field at the top of the page.

2. Tap a search suggestion, or tap Go on the keyboard to search for exactly what you typed.

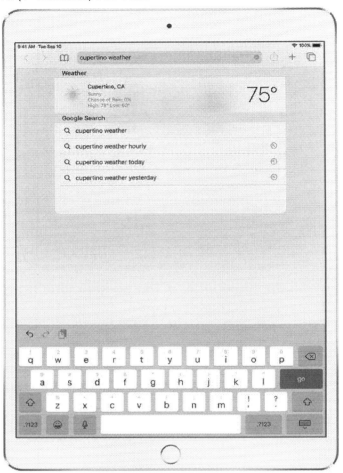

If you don't want to see suggested search terms, go to Settings 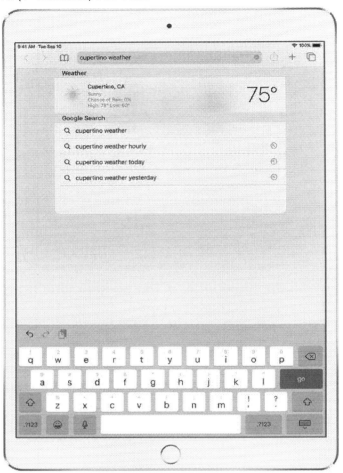 > Safari, then turn off Search Engine Suggestions (below Search).

Find websites you visited before

Safari search suggestions include your open tabs, bookmarks, and pages you recently visited. For example, if you search for "iPad," the search suggestions under Switch to Tab include your open tabs related to "iPad." Websites related to "iPad" that you bookmarked or recently visited appear under Bookmarks and History.

Search within websites

To search within a website, enter a website followed by a search term in the search field. For example, enter "wiki einstein" to search Wikipedia for "einstein."

To turn this feature on or off, go to Settings > ⚙ > Safari > Quick Website Search.

See your favorites when you enter an address, search, or create a new tab

Go to Settings ⚙ > Safari > Favorites, then select the folder with the favorites you want to see.

Search the page

You can find a specific word or phrase on a page.

1. Tap ⬆, then tap Find on Page.

2. Enter the word or phrase in the search field.

3. Tap ⌄ to find other instances.

Choose a search engine

Go to Settings ⚙ > Safari > Search Engine.

Use Shortcuts to automate tasks on iPad

The Shortcuts app 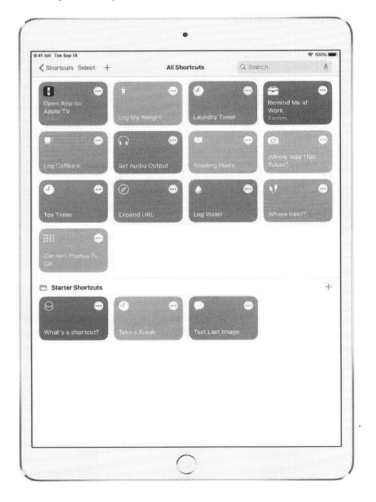 lets you automate tasks you do often with just a tap or by asking Siri. Create shortcuts to get directions to the next event in your Calendar, move text from one app to another, and more. Choose ready-made shortcuts from the Gallery or build your own using different apps to run multiple steps in a task.

Check stocks on iPad

Use the Stocks app  on iPad to track market activity, view the performance of stocks you follow, and get the latest business news.

Ask Siri. Say something like: "How are the markets doing?" or "How's Apple stock today?"

Manage your watchlist

Add the stocks you follow to your watchlist to quickly view price, price change, percentage change, and market capitalization values.

- *Add a symbol to your watchlist:* In the search field, enter a stock symbol, company name, fund name, or index. Tap the symbol you want to add in the search results, then tap Add to Watchlist.

- *Delete a symbol:* Swipe left on the symbol in your watchlist, then tap Remove.

- *Reorder symbols:* Tap Edit at the top of the watchlist. Touch and hold ☰ to drag a symbol up or down, then tap Done.

View stock charts, details, and news

Tap a stock symbol in your watchlist to view an interactive chart, additional details, and related news stories.

- *See the performance of a stock over time:* Tap an option from the time range selections at the top of the chart.

- *See the value for a specific date or time:* Touch and hold the chart with one finger.

- *See the difference in value over time:* Touch and hold the chart with two fingers.

- *See more details:* Below the chart, view additional stock details like 52-week high and low, Beta, EPS, and average trading volume.

- *Read news:* Swipe up to see additional news stories, then tap a story.

View your watchlist across devices

You can view your watchlist on your iPhone, iPad, iPod touch, and Mac when you're signed in with the same Apple ID.

On your iPhone, iPad, and iPod touch, go to Settings > [*your name*] > iCloud, then turn on Stocks.

On your Mac, choose Apple menu > System Preferences, then do one of the following:

- *macOS 10.15 or later:* Click Apple ID, select iCloud, then turn on Stocks.

- *macOS 10.14 or earlier:* Select iCloud, then turn on Stocks.

Read business news

Tap Business News above the watchlist, then swipe up on Top Stories to view stories selected by Apple News editors that highlight the current news driving the market. You also see stories about companies in your watchlist grouped by ticker symbol, including Apple News+ content for subscribers (Apple News and Apple News+ are not available in all countries or regions).

Stories from publications you have blocked in Apple News don't appear in the news feed.

Add a Stocks widget to your iPad Home Screen

Add a Stocks widget to Today View on your iPad Home Screen to check stocks at a glance. Choose Watchlist to monitor several symbols from your watchlist on your Home Screen, or Symbol to monitor the performance of a single symbol.

Get tips on iPad

In the Tips app 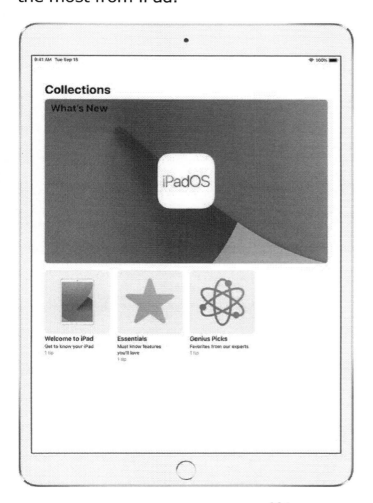, see collections of tips that help you get the most from iPad.

Get Tips

In the Tips app, tap a collection to learn how to take better photos, use dictation, create a custom radio station, and much more.

New tips are added frequently.

Get notified when new tips arrive

1. Go to Settings , then tap Notifications.

2. Tap Tips below Notification Style, then turn on Allow Notifications.

3. Choose options for the location and style of tip notifications, when they should appear, and so on.

Set up the Apple TV app on iPad

With the Apple TV app , you can watch Apple TV+ originals as well as your other favorite shows, movies, sports, and live news. Subscribe to Apple TV channels such as CBS All Access and Showtime, watch titles from streaming services and cable providers, and purchase or rent movies and TV shows. The Apple TV app is on your iPhone, iPad, iPod touch, Mac, Apple TV, and supported smart TVs and streaming devices, so you can watch at home or wherever you go.

Note: The availability of the Apple TV app and its features and services (such as Apple TV+, Apple TV channels, sports, news, and supported apps) varies by country or region.

Subscribe to Apple TV+

You can subscribe to Apple TV+ (not available in all countries or regions) and watch new, exclusive Apple Originals without ads. Stream Apple TV+ on demand on iPhone, iPad, iPod touch, Mac, Apple TV, and supported smart TVs and streaming devices, or download Apple Originals to watch offline on iPhone, iPad, iPod touch, and Mac. If you use Family Sharing, up to five other family members can share the subscription for no additional charge.

1. Tap Watch Now, scroll down to the Apple TV+ row, then do one of the following:

   - *Start a free trial:* Tap the button to start your free trial (available for eligible Apple ID accounts). Apple TV+ provides one free trial per subscriber or family.

   - *Start a monthly subscription:* Tap Subscribe.

   Review the subscription details, then confirm with Face ID, Touch ID, or your Apple ID.

Subscribe to Apple TV channels

If you subscribe to Apple TV channels (such as CBS All Access and Showtime), you can stream the ad-free content on demand or download it to watch offline. If you use Family Sharing, up to five other family members can share the subscription for no additional charge.

1. Tap Watch Now, then scroll down to browse the available channels.

2. To watch a channel, do one of the following:

   - *Start a free trial:* Tap the button to start your free trial (available for eligible Apple ID accounts). Each Apple TV channel provides one free trial per subscriber or family. The length of the trial may vary.

   - *Start a monthly subscription:* Tap Subscribe.

   Review the subscription details, then confirm with Face ID, Touch ID, or your Apple ID.

Add your cable or satellite service to the Apple TV app

Single sign-on provides immediate access to all the supported video apps in your subscription package.

1. Go to Settings > TV Provider.

2. Choose your TV provider, then sign in with your provider credentials.

If your TV provider isn't listed, sign in directly from the app you want to use.

Connect supported apps to the Apple TV app

The Apple TV app recommends new content or the next episode in a series you watched. The first time you play from a supported app, tap Connect to allow the connection to the Apple TV app.

Manage your connected apps and subscriptions

1. Tap Watch Now, then tap ⊙ or your profile picture at the top right.

2. Tap any of the following:

   - *Connected Apps:* Turn apps on or off.

     Connected apps appear in the Apple TV app on all your devices where you're signed in with your Apple ID.

   - *Manage Subscriptions:* Tap a subscription to change or cancel it.

   - *Clear Play History:* Remove your viewing history from all your Apple devices.

Find shows, movies, and more in the Apple TV app on iPad

The Apple TV app brings your favorite shows, movies, sports, and live news together in one place. Quickly find and watch your favorites, pick up where you left off with Up Next, or discover something new—personalized just for you.

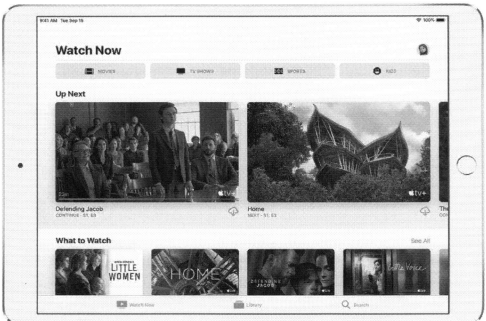

Watch *Defending Jacob* on the Apple TV app; *Home* is also available on the Apple TV app

Discover shows and movies

In the Apple TV app, tap Watch Now, then do any of the following:

- *See what's up next:* In the Up Next row, find titles you recently added, rented, or purchased, catch the next episode in a series you've been watching, or resume what you're watching from the moment you left off.

- *Browse by category:* Tap a category such as Movies, TV Shows, Sports, or Kids.

- *Get recommendations:* Browse the What to Watch row for editorial recommendations that are personalized for you.

Many rows throughout the app feature personalized recommendations based on your channel subscriptions, supported apps, purchases, and viewing interests.

- *Browse Apple TV+:* In the Apple TV+ row, tap a title to see more information or play a trailer.

- *Browse Apple TV channels:* Scroll down to browse channels you subscribe to. In the Apple TV channels row, browse other available channels, then tap a channel to explore its titles.

- *Watch live news:* (available in select countries or regions) Tap TV shows at the top, scroll down to the News row, then choose a news channel.

Search for a title, sport, team, cast member, or Apple TV channel

Tap Search, then enter what you're looking for in the search field.

Explore sports (U.S. and Canada only)

You can browse by sport or league, watch games, and get live scores and notifications for close games.

1. Tap Watch Now, then tap Sports at the top.

2. Do any of the following:

   - *Browse by sport:* Scroll down to browse the events of many sports, such as football, baseball, and basketball.

To narrow your browsing, scroll down, then choose a sport.

- *Watch a game:* Tap the game.

- *Choose your favorite teams:* Scroll to the bottom, then tap Your Favorite Teams.

    Their games automatically appear in Up Next, and you'll receive notifications about your favorite teams.

To hide the scores of live games, go to Settings ⚙ > TV, then turn off Show Sports Scores.

Use the Up Next queue

- *Add an item to Up Next:* Tap the item to see its details, then tap Add.

- *Remove an item from Up Next:* Touch and hold the item, then tap Remove from Up Next.

- *Continue watching on another device:* You can see your Up Next queue in Watch Now on your iPhone, iPad, iPod touch, Mac, Apple TV, or supported smart TV where you're signed in with your Apple ID.

Watch shows and movies in the Apple TV app on iPad

Play shows and movies from the Apple TV app on iPad. Purchases, rentals, Apple TV+, and Apple TV channels play in the Apple TV app, while content from other providers plays in their video app.

Buy, rent, stream, or download shows and movies

1. Tap an item to see its details.

2. Choose any of the following options (not all options are available for all titles):

   - *Watch Apple TV+ or Apple TV channels:* Tap Play. If you're not a current subscriber, tap Try It Free (available for eligible Apple ID accounts) or Subscribe.

   - *Choose a different video app:* If the title is available from multiple apps, scroll down to How To Watch, then choose an app.

   - *Buy or rent:* Confirm your selection, then complete the payment.

     When you rent a movie, you have 30 days to start watching it. After you start watching the movie, you can play it as many times as you want for 48 hours, after which the rental period ends. When the rental period ends, the movie is deleted.

   - *Download:* Tap ☁⬇. You can find the download in Library and watch it even when iPad isn't connected to the internet.

   - *Pre-order:* Review the details, then tap Pre-Order.

242

When the pre-ordered item becomes available, your payment method is billed, and you receive an email notification. If you turned on automatic downloads, the item is automatically downloaded to your iPad.

Note: The availability of Apple Media Services varies by country or region.

Control playback in the Apple TV app

During playback, tap the screen to show the controls.

The LEGO Movie 2: The Second Part is available on the Apple TV app

Control   Description

Control Description

▶   Play

❚❚   Pause

⟲15   Skip backward 15 seconds; touch and hold to
         rewind

⟳15   Skip forward 15 seconds; touch and hold to fast-
         forward

▢   Change the aspect ratio; if you don't see the
         scaling control, the video already fits the screen
         perfectly

🖵   Display subtitles and closed captions (if available)

🖵   Stream the video to other devices

🖵   Multitask with Picture in Picture; you can continue
         to watch the video while you use another app

Control   Description

✕   Stop playback

Change the Apple TV app settings

1. Go to Settings 🍥 > TV.

2. Choose streaming options:

- *Use Cellular Data:* (Wi-Fi + Cellular models) Turn off to limit streaming to Wi-Fi connections.

- *Cellular:* (Wi-Fi + Cellular models) Choose High Quality or Automatic.

- *Wi-Fi:* Choose High Quality or Data Saver.

  High Quality requires a faster internet connection and uses more data.

  Choose download options:

- *Use Cellular Data:* (Wi-Fi + Cellular models) Turn off to limit downloads to Wi-Fi connections.

- *Cellular:* (Wi-Fi + Cellular models) Choose High Quality or Fast Downloads.

- *Wi-Fi:* Choose High Quality or Fast Downloads.

  High Quality results in slower downloads and uses more data.

- *Languages:* Choose a language. Each added audio language increases the download size. To remove a language, swipe left on the language you want to remove, then tap Delete.

  The default language is the primary language for your country or region. If you turned on Audio Descriptions in Settings > Accessibility, audio descriptions are also downloaded.

To update your recommendations and Up Next queue based on what you watch on your iPad, turn on Use Play History.

What you watch on your iPad affects your personalized recommendations and Up Next on all your devices where you're signed in with your Apple ID.

Remove a download

1. Tap Library, then tap Downloaded.

2. Tap Edit, select the item you want to remove, then tap Delete.

Removing an item from iPad doesn't delete it from your purchases in iCloud. You can download the item to iPad again later.

Make a recording in Voice Memos on iPad

With the Voice Memos app 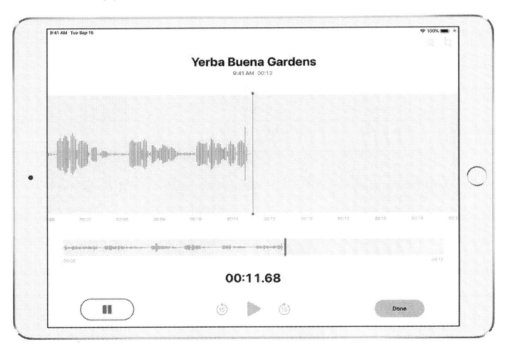, you can use iPad as a portable recording device to record personal notes, classroom lectures, musical ideas, and more. You can fine-tune your recordings with editing tools like trim, replace, and resume.

Record voice memos using the built-in microphone, a supported headset, or an external microphone.

When Voice Memos is turned on in iCloud settings or preferences, your recordings appear and update automatically on all your devices where you're signed in with the same Apple ID.

Make a basic recording

1. To begin recording, tap .

   To adjust the recording level, move the microphone closer to or farther from what you're recording.

2. Tap Done to finish recording.

Your recording is saved with the name New Recording or the name of your location, if Location Services is turned on in Settings > Privacy. To change the name, tap the recording, then tap the name and type a new one.

To fine-tune your recording, see Edit a recording in Voice Memos.

Note: For your privacy, when you use Voice Memos to make recordings, an orange dot appears at the top of your screen to indicate your microphone is in use.

Use the advanced recording features

You can make a recording in parts, pausing and resuming as you record.

1. To begin recording, tap .

   To adjust the recording level, move the microphone closer to or farther away from what you're recording.

To see more details while you're recording, swipe up from the top of the waveform.

2. Tap ❚❚ to stop recording; tap Resume to continue.

3. To review your recording, tap ▶.

   To change where playback begins, drag the playhead right or left across the small waveform at the bottom of the screen before you tap ▶.

4. To save the recording, tap Done.

Your recording is saved with the name New Recording or the name of your location, if Location Services is turned on in Settings ⚙ > Privacy. To change the name, tap the recording, then tap the name and type a new one.

To fine-tune your recording, see Edit a recording in Voice Memos.

Mute the start and stop tones

While recording, use the iPad volume down button to turn the volume all the way down.

Use another app while recording

While you're recording, you can use another app, as long as it doesn't play audio on your device. If the app starts playing or recording sound, Voice Memos stops recording.

1. While recording, you can go to the Home Screen and open another app.

2. To return to Voice Memos, tap the red bar at the top of the screen.

If Voice Memos is turned on in iCloud settings or preferences, your recording is saved in iCloud and appears automatically on all your devices where you're signed in with the same Apple ID.

Recordings using the built-in microphone are mono, but you can record stereo using an external stereo microphone that works with iPad. Look for accessories marked with the Apple "Made for iPad" or "Works with iPad" logo.

Chapter 6

Siri

How to Ask Siri on iPad

Talking to Siri is a quick way to get things done. Ask Siri to translate a phrase, set a timer, find a location, report on the weather, and more. The more you use Siri, the better it knows what you need.

To use Siri, iPad must be connected to the Internet. Cellular charges may apply.

Set up Siri

If you didn't set up Siri when you first set up your iPad, go to Settings > Siri & Search, then turn on the following:

- Listen for "Hey Siri"

- Press Home for Siri (models with the Home button) or Press Top Button for Siri (other models)

Summon Siri with your voice

- Say "Hey Siri," then ask Siri a question or to do a task for you.

- For example, say something like "Hey Siri, how's the weather today?" or "Hey Siri, set an alarm for 8 a.m."

- To ask Siri another question or to do another task, tap the Listen button.

Note: To prevent iPad from responding to "Hey Siri," place your iPad face down, or go to Settings > Siri & Search, then turn off Listen for "Hey Siri."

Summon Siri with a button

Do one of the following:

- Models with the Home button: Press and hold the Home button.

- Other models: Press and hold the top button.

- When Siri appears, ask Siri a question or to do a task for you.

- For example, say something like "What's 18 percent of 225?" or "Set the timer for 3 minutes."

- To ask Siri another question or to do another task, tap the Listen button.

Make a correction if Siri misunderstands you

- Rephrase your request: Tap the Listen button, then say your request in a different way.

- Spell out part of your request: Tap the Listen button, then repeat your request by spelling out any words that Siri didn't understand. For example, say "Call," then spell the person's name.

- Edit your request with text: Above the response from Siri, tap "Tap to Edit," then use the onscreen keyboard.

- Change a message before sending it: Say "Change it."

Type instead of speaking to Siri

- Go to Settings > Accessibility > Siri, then turn on Type to Siri.

- To make a request, summon Siri, then use the keyboard and text field to ask Siri a question or to do a task for you.

Siri is designed to protect your information, and you can choose what you share. To learn more, go to Settings > Siri & Search > About Ask Siri & Privacy.

Find out what Siri can do on iPad

Use Siri on iPad to get information and perform tasks.

- **Find answers to your questions:** Find information from the web, get sports scores, get arithmetic calculations, and more. Say something like "Hey Siri, what causes a rainbow," "Hey Siri, what was the score of the Orioles game yesterday," or "Hey Siri, what's the derivative of cosine x?"

When Siri displays a web link, you can tap it to see more information in Safari.

- **Perform tasks with apps on iPad**: Use Siri to control apps with your voice. For example, to create an event in Calendar, say something like "Hey Siri, set up a meeting with Gordon at 9," or to add an item to Reminders, say something like "Hey Siri, add artichokes to my groceries list."

When the onscreen response from Siri includes buttons or controls, you can tap them to take further action.

- **Translate languages**: Say something like "Hey Siri, how do you say Thank You in Mandarin?" or "Hey Siri, what languages can you translate?"

In response to the question "How do you say thank you in Mandarin?," Siri displays a translation of the English phrase "thank you" into Mandarin. A button to the right of the translation replays audio of the translation.

- **Play a radio station**: Say something like "Hey Siri, play Wild 94.9" or "Hey Siri, tune into ESPN Radio."

- Let Siri show you more examples: Say something like "Hey Siri, what can you do?" You can also tap the Help button after you summon Siri.

About Siri Suggestions on iPad

Siri makes suggestions for what you might want to do next, such as confirm an appointment or send an email, based on your routines and how you use your apps. For example, Siri might help when you do any of the following:

- **Glance at the Lock screen or start a search**: As Siri learns your routines, you get suggestions for just what you need, at just the right time. For example, if you frequently order coffee mid morning, Siri may suggest your order near the time you normally place it.

- **Create email and events**: When you start adding people to an email or calendar event, Siri suggests the people you included in previous emails or events.

- **Leave for an event**: If your calendar event includes a location, Siri assesses traffic conditions and notifies you when to leave.

- **See your flight status**: If you have a boarding pass in Mail, Siri shows your flight status in Maps. You can tap the suggestion when you're ready to get directions to the airport.

- **Type**: As you enter text, Siri can suggest names of movies, places—anything you viewed on iPad recently. If you tell a friend you're on your way, Siri can even suggest your estimated arrival time.

- **Search in Safari**: Siri suggests websites and other information in the search field as you type.

- (iPad Air 2 and later) Above the keyboard, Siri also suggests words and phrases based on what you were just reading.

- **Confirm an appointment or book a flight on a travel website:** (iPad Air 2 and later) Siri asks if you want to add it to your calendar.

- **Read News stories**: As Siri learns which topics you're interested in, they'll be suggested in News.

To turn off Siri Suggestions, go to Settings > Siri & Search, then turn off any of the following:

- Suggestions in Search

- Suggestions in Lookup

- Suggestions on Lock Screen

For a specific app, tap the app, then turn off Show Siri Suggestions.

Your personal information—which is encrypted and remains private—stays up to date across all your devices where you're signed in with the same Apple ID. As Siri learns about you on one device, your experience with Siri is improved on your other devices.

Siri is designed to protect your information, and you can choose what you share. To learn more, go to Settings > Siri & Search > About Search Suggestions & Privacy.

Change Siri settings on iPad

You can change the voice for Siri, prevent access to Siri when your device is locked, and more. Go to Settings > Siri & Search, then do any of the following:

- **Change the voice for Siri**: (not available in all languages) Tap Siri Voice, then choose a male or female voice for Siri or change the accent.

- Prevent Siri from responding to the voice command "Hey Siri": Turn off Listen for "Hey Siri."

- Prevent Siri from responding to the Home button or the top button: Turn off Press Home for Siri (models with the Home button) or Press Top Button for Siri (other models).

- Change the language Siri responds to: Tap Language.

- Limit when Siri provides voice feedback: If you don't want Siri to always provide voice feedback, tap Voice Feedback, then choose an option.

- Prevent access to Siri when iPad is locked: Turn off Allow Siri When Locked.

Adjust the Siri voice volume

Ask Siri. Say something like: "Turn up the volume" or "Turn down the volume."

Chapter 7

Apple Pay

Set up Apple Pay

Set up Apple Pay to make secure payments in apps and on websites that support Apple Pay. In Messages, you can send and receive money from friends and family or make purchases using Business Chat.

Add a credit or debit card

- Go to Settings > Wallet & Apple Pay.

- Tap Add Cards. You may be asked to sign in with your Apple ID.

Do one of the following:

- **Add a new card**: Position iPad so that your card appears in the frame, or enter the card details manually.

- **Add your previous cards**: Select the card associated with your Apple ID, cards you use with Apple Pay on your other devices, or cards that you removed. Tap Continue, then enter the CVV number of each card.

Alternatively, you may be able to add your card from the app of the bank or card issuer.

The card issuer determines whether your card is eligible for Apple Pay, and may ask you for additional information to complete the verification process.

View the information for a card and change its settings

- Go to Settings > Wallet & Apple Pay.

- Tap a card, then do any of the following:

- Tap Transactions to view your recent history. To hide this information, turn off Transaction History. To view all your Apple Pay activity, see the statement from your card issuer.

- View the last four digits of the card number and Device Account Number—the number transmitted to the merchant.

- Change the billing address.

- Remove the card from Apple Pay.

Change your Apple Pay settings

- Go to Settings > Wallet & Apple Pay.

- Do any of the following:

- Set your default card.

- Add the shipping address and contact information for purchases.

Remove your cards from Apple Pay if your iPad is lost or stolen

If you enabled Find My iPhone, use it to help locate and secure your iPad. Do any of the following:

- On a Mac or PC: Sign in to your Apple ID account. In the Devices section, click the lost iPad. Below the list of cards, click Remove all.

- On another iPhone, iPad, or iPod touch: Go to Settings > [your name], tap the lost iPad, then tap Remove All Cards (below Apple Pay).

- Call the issuers of your cards.

If you remove cards, you can add them again later. If you sign out of iCloud in Settings > [your name], all your credit and debit cards for Apple Pay are removed from iPad. You can add the cards again the next time you sign in.

Pay in apps or on the web using Apple Pay on iPad

Use Apple Pay to make purchases in apps and on the web in Safari wherever you see the Apple Pay button.

Pay in an app or on the web

- During checkout, tap the Apple Pay button.

- Review the payment information.

- You can change the credit card, shipping address, and contact information.

- Complete the payment:

- Models with Face ID: Double-click the top button, then glance at iPad to authenticate with Face ID, or enter your passcode.

- Models with Touch ID: Authenticate with Touch ID or enter your passcode.

Change your default shipping and contact information

- Go to Settings > Wallet & Apple Pay.

- Set any of the following:

- Shipping address

- Email

- Phone

Set up and use Apple Cash on iPad (U.S. only)

When you receive money in Messages, it's added to your Apple Cash. You can use Apple Cash right away wherever you would use Apple Pay. You can also transfer your Apple Cash balance to your bank account.

Set up Apple Cash

Do any of the following:

- Go to Settings > Wallet & Apple Pay, then turn on Apple Cash.

- In Messages, send or accept a payment. See Pay in apps or on the web using Apple Pay.

Use Apple Cash

You can use Apple Cash wherever you use Apple Pay:

- Send and receive money with Apple Pay (U.S. only)

- Pay in apps or on the web using Apple Pay

- Manage your Apple Cash

Go to Settings > Wallet & Apple Pay, then tap Apple Cash. Do any of the following:

- Add money from a debit card.

- Transfer money to your bank.

- Update your bank account information.

- Tap Transactions to view your history and details (including comments sent with payments), manually accept or reject individual payments, and request a statement.

- Choose to manually or automatically accept all payments. You have 7 days to manually accept a payment before it's returned to the sender.

- Verify your identity for account servicing and to increase your transaction limits.

- Contact Apple Support.

Transfer money from Apple Cash to your Visa debit card or bank account

You can transfer money from your Apple Cash1 balance instantly or within 1 to 3 business days.

You can use Instant Transfer to transfer money from your Apple Cash balance to an eligible Visa debit card, or you can use a bank transfer to transfer money to your bank account.

To use Instant Transfer, your iPhone needs to have iOS 12.2 or later. If you are transferring money to a debit card with Instant Transfer, it must be an eligible Visa debit card.

Use Instant Transfer

With Instant Transfer, you can quickly transfer money to an eligible Visa debit card in Wallet. An Instant Transfer is typically available within 30 minutes. Go to your card info:

- **iPhone**: open the Wallet app, tap your Apple Cash card, then tap the more button.

- **iPad**: open the Settings app, tap Wallet & Apple Pay, then tap your Apple Cash card.

- Tap Transfer to Bank.

- Enter an amount and tap Next.

- Tap Instant Transfer.

- If you haven't added a Visa debit card, tap Add Card and follow the instructions on your screen to add one.

- Tap > select the Visa debit card you want to transfer funds to and select the billing information for your chosen debit card.

- Your funds should transfer within 30 minutes.

Transfer in 1 to 3 business days

You can also use a bank transfer to transfer money to your bank account within 1 to 3 business days.

- Go to your card info:

- iPhone: Open the Wallet app, tap your Apple Cash card, then tap the more button.

- iPad: Open the Settings app, tap Wallet & Apple Pay, then tap your Apple Cash card.

- Apple Watch: Transfer money using your iPhone.

- Tap Transfer to Bank.

- Enter an amount and tap Next.

- Choose 1-3 Business Days. If you don't have a bank account set up, follow the instructions on your screen to add one.

- Confirm with Face ID, Touch ID, or passcode.

- Wait for the money to transfer. This can take 1 to 3 business days.

How long does a bank transfer take

If you use Instant Transfer, your funds are typically available in your bank account within 30 minutes.

Bank transfers usually take 1 to 3 business days to complete. Check your bank statement to see if the transfer has been processed and deposited into your bank account. Bank transfers aren't deposited on bank holidays or weekends. See holidays observed by the Federal Reserve on federalreserve.gov.

How to update your bank account information

- Go to your card info:

- iPhone: Open the Wallet app, tap your Apple Cash card, then tap the more button.

- iPad: Open the Settings app, tap Wallet & Apple Pay, then tap your Apple Cash card.

- Apple Watch: Edit the information using your iPhone.

- Tap Bank Account, then tap the bank account you want to update.

- To delete your banking information, tap Delete Bank Account Information. Tap again to confirm. After deleting, you can add your bank information again or add different information.

- To edit your bank information, tap next to your routing number or account number, add your information, then confirm the numbers and tap Next.

After you update your information on one device, it automatically updates on all the devices where you're signed in with your Apple ID.

Chapter 8
Family Sharing

Set up Family Sharing on iPad

With Family Sharing, up to six family members can share iTunes Store, App Store, and Apple Books purchases; an Apple Music family membership; an Apple News+ subscription; an Apple Arcade subscription; an iCloud storage plan; Screen Time information; a family calendar; family photos; and more, all without sharing accounts.

To use Family Sharing, one adult family member (the *organizer*) chooses features for the family to share and invites up to five additional family members to participate. When family members join, Family Sharing is set up on their devices automatically.

Family Sharing requires you (the organizer) to sign in with your Apple ID and to confirm the Apple ID you use for the iTunes Store, the App Store, and Apple Books (you usually use the same Apple ID for everything). Family Sharing is available on devices that meet these minimum system requirements: iOS 8, iPadOS 13, a Mac with OS X 10.10, or a PC with iCloud for Windows 7. You can be part of only one family group at a time.

Note: You can set up Screen Time for individual family members either through Family Sharing on your device or separately on their own devices.

Get started with Family Sharing:

1. Go to Settings ⚙ > [*your name*] > Set Up Family Sharing, then follow the onscreen instructions.

2. Tap the features you want to share:

   - Purchase Sharing
   - iCloud Storage
   - Location Sharing
   - Screen Time
   - Apple Music
   - TV Channels
   - Apple Arcade
   - Apple News+

   Follow the onscreen instructions to sign up.

Depending on the features you choose, you may be asked to set up an Apple Music family membership or an iCloud Storage subscription. If you choose to share iTunes Store, App Store, and Apple Books purchases with your family members, you agree to pay for any purchases they initiate while part of the family group.

Create an Apple ID for a child

1. Go to Settings ⚙ > [*your name*] > Family Sharing > Add Family Member.

2. Tap Create a Child Account, then follow the onscreen instructions.

   The child's account is added to your family until the child is at least 13 years old.

Accept an invitation to Family Sharing

- Tap Accept in the invitation.

Or, if you're near the organizer during the setup process, you can enter your Apple ID and password on the Family Member's Apple ID screen on the organizer's device.

Leave Family Sharing

Any family member can leave the Family Sharing group, but only the organizer can stop family sharing.

1. Go to Settings ⚙ > [*your name*] > Family Sharing > [*your name*].

2. Tap Leave Family.

   If you're the organizer, tap Stop Family Sharing.

Share purchases with family members on iPad

With Family Sharing, up to six family members can share iTunes Store, App Store, and Apple Books purchases, an Apple Music family membership, and an iCloud storage plan.

When your family shares iTunes Store, App Store, and Apple Books purchases, all items are billed directly to the family organizer's Apple ID account. Once purchased, an item is added to the initiating family member's account and eligible purchases are shared with the rest of the family.

Access shared purchases from the iTunes Store

1. Open the iTunes Store ⊠, then tap Purchased.

2. Tap My Purchases at the top left.

3. Choose a family member.

4. Tap a category (for example, Music or Movies) at the top of the screen, tap a purchased item, then tap ⬇ to download it.

Access shared purchases from the App Store

1. Open the App Store ⬛.

2. Tap ⬤—or your profile picture—at the top right.

3. Tap Purchased, choose a family member, then tap ⬇ next to a purchased item to download it.

Access shared purchases from Apple Books

1. Open the Books app 📖.

2. Tap ⊙, or your profile picture, at the top right.

3. Tap an item under My Purchases or choose a family member, then tap a category (for example, Books or Audiobooks).

4. Tap All Books, Recent Purchases, or a genre, then tap ⬇ next to a purchased item to download it.

Use a shared iCloud storage plan

With Family Sharing, your family can share an iCloud storage plan of 200 GB or 2 TB.

1. Go to Settings ⚙ > [your name] > Family Sharing.

2. Tap iCloud Storage, then follow the onscreen instructions.

You can also purchase or use your own storage plan if you need more space. To upgrade your iCloud storage.

Turn on Ask To Buy on iPad

When your Family Sharing group shares purchases, you can, as the family organizer, require that young family members request approval for purchases or free downloads.

1. Go to Settings ⚙ > [your name] > Family Sharing.

2. Tap the name of the person who needs to request approval, then turn on Ask To Buy.

Note: Age restrictions for Ask To Buy vary by region. In the United States, the family organizer can turn on Ask To Buy for any family member under age 18; for children under age 13, it's turned on by default.

Hide your purchases on iPad

With Family Sharing, you can hide your iTunes Store, App Store, and Apple Books purchases from family members.

1. Go to Settings ⚙ > [*your name*] > Family Sharing.

2. Tap Purchase Sharing, then turn off Share My Purchases.

Share subscriptions with family members on iPad

When you're in a Family Sharing group, you can share a subscription to Apple Music and Apple News+ with family members.

Use a shared Apple Music family membership

If your family has an Apple Music family membership, you can simply open Apple Music and start listening. If you don't have an Apple Music family membership, you can sign up for one.

1. Go to Settings ⚙ > [*your name*] > Family Sharing.

2. Tap Apple Music, then follow the onscreen instructions.

Each family member gets their own music library and personal recommendations. (Selections are subject to content restrictions set in Screen Time.) To listen to music, make sure you're signed in with the Apple ID you entered in Family Sharing settings.

Use a shared Apple News+ subscription

In News, you can subscribe to Apple News+, which includes over 200 magazines and other publications. When you're in a Family Sharing group and you buy an Apple News+ subscription, all members of your family can read Apple News+ publications for no additional charge. The other members of your family get a message about Apple News+ when they open News. If you don't have a subscription, you can sign up through the News app (not available in all countries or regions).

Use a shared Apple Arcade subscription

Your family can share a subscription to Apple Arcade (not available in all countries or regions), a service that gives you access to new games without ads or additional purchases. With an Apple Arcade subscription, all members of your family (up to six people) can download and play Apple Arcade games from the App Store. (Selections are subject to content restrictions set in Screen Time.) The subscription allows you to play across iPhone, iPad, iPod touch, Mac, and Apple TV. Each player gets a personalized account—your progress is carried over between devices.

New games are added to Apple Arcade regularly. If you don't have a subscription, you can sign up for one through the App Store (not available in all countries or regions).

Share photos, a calendar, and more with family members on iPad

With Family Sharing on iPad, you can share a family photo album, a family calendar, your location, and more.

Share photos or videos with family members

When you set up Family Sharing, a shared album called Family is automatically created in the Photos app on all family members' devices, which makes it easy to share family photos or videos.

1. Open the Photos app ⚜, then select a photo or video, or select multiple photos or videos.

2. Tap ⬆, then tap Shared Albums.

3. Add any comments you want, then tap Shared Albums.

4. Choose an album to add the photo or video to.

Members can unsubscribe from the family album, and the family organizer can delete it or remove subscribers.

Add an event to the family calendar

When you set up Family Sharing, a shared calendar called Family is automatically created in the Calendar app on all family members' devices. You can schedule an event on this calendar to share it with family members.

1. Open the Calendar app ⌞10⌟, then add an event.

2. While entering the event details, tap Calendar, then tap Family to add the event to the family calendar.

Members can unsubscribe from the family calendar, and the family organizer can delete it.

Share your location with family members

With Family Sharing, you can share your location with members of your family group. When the family organizer sets up Location Sharing in Family Sharing settings, the organizer's location is automatically shared with everyone in the family. Then family members can choose whether or not to share their location.

When you share your location, your family members can see your location in **Find My** on and in Messages and, if your device is lost, they can help you locate it with **Find My** on.

Note: To share your location, you must have Location Services turned on in Settings 🌐 > Privacy.

1. Go to Settings > [*your name*] > Family Sharing > Location Sharing, then turn on Share My Location.

2. Tap Change My Location to This iPad.

3. Tap a family member you want to share your location with, tap Share My Location, then tap ⟨.

You can repeat this step for each family member you want to share your location with. Each family member receives a message that you're sharing your location and can choose to share their location with you.

You can also send or share your location from the Messages app (iOS 8, iPadOS 13, or later) by tapping the profile picture or name at the top of the conversation, tapping 🛈, then tapping Send My Current Location or Share My Location.

To stop sharing your location with a family member, tap the profile picture or name at the top of the conversation with the family member, tap 🛈, then tap Stop Sharing My Location.

To find a family member's location, use the **Find My** on app .

Share a Personal Hotspot

With Family Sharing, you can share an Internet connection through a Personal Hotspot with members of your family group. When a member of your family group sets up a Personal Hotspot, other family members can use it without having to enter the password..

Locate a family member's missing device on iPad

When you're in a Family Sharing group and family members share their locations with you, you can use the Find My app ⚙ on your iPhone, iPad, or iPod touch, or on iCloud.com on a computer to help a family member find a lost device.

Set up your device to be found by a family member

A family member can help locate your missing device if you do the following on the device *before* it's lost:

- *Turn on Location Services:* Go to Settings ⚙ > Privacy, then turn on Location Services.

- *Turn on Find My iPad:* Go to Settings > [*your name*] > Find My > Find My iPad, then turn on Find My iPad, Enable Offline Finding, and Send Last Location.

- *Share your location with family members:* Go to Settings > [*your name*] > Family Sharing > Location Sharing, then turn on Share My Location and make sure Find My iPad is turned on.

Locate a family member's device

1. Open Find My on your iPhone, iPad, or iPod touch, or open Find My iPhone on iCloud.com on a computer.

2. Sign in with your **Apple ID**.

3. In the list of devices, select the one you want to find.

Your devices are at the top of the list, and your family members' devices are below yours.

The selected device appears on a map so you can see where it is.

Set up Screen Time for family members on iPad

You can set up Screen Time for family members through Family Sharing—including downtime, allowances for app use, the contacts your family communicates with, content ratings, and more. Screen Time also lets you and your family members see how they're using their devices and use that information to structure their device use.

You can invite family members to join and enter their Apple IDs in Family Sharing on your device to begin the process. Or you can set up Screen Time for them in Family Sharing on your device.

Note: When you set up Screen Time for a family member through Family Sharing, notifications of your family member's weekly report appear on both your device and your family member's. To view the report, tap the notification.

1. Go to Settings ⚙ > [*your name*] > Family Sharing > Screen Time.

2. Tap a family member, then tap Turn on Screen Time.

3. Tap Continue, then follow the onscreen instructions.

Important: If you set up Screen Time for a family member through Family Sharing and you forget the family Screen Time passcode, you can reset it on your device using your device passcode, Touch ID, or Face ID.

Chapter 9

Screen Time

View your Screen Time summary on iPad

Screen Time shows you how you use your iPhone, iPad, iPod touch, and Mac—including which apps and websites you spend time with, how often you pick up your iPad, and so on. You can use this information to help you make decisions about managing the time you spend on your devices. You can set allowances and limits for using certain apps and websites, prevent access to explicit music and web content, and more.

When you set up Screen Time, it begins building a description of your device use, including:

- How much time you spend using apps by category (social networking, entertainment, reading, and so on)

- A breakdown of your app use by time of day

- How long you spend using each app, and which apps you used beyond your time limit

- An overview of the types of notifications you get, and which apps are sending you the most notifications

- How often you pick up your device and which apps you use—that is, how many times each app was the first one used after picking up the device. You can tap each app in your

Screen Time summary to see more information about its use.

When Screen Time is set up, you can view your summary in

Settings  > Screen Time > See All Activity. You can see a

summary of your device use for the current day or the past

week.

Set up Screen Time for yourself on iPad

With Screen Time, you can set allowances and limits for your app use, schedule downtime, and more. You can change or turn off any of these settings at any time.

Set downtime

You can block apps and notifications during periods when you want time away from your devices.

1. Go to Settings > Screen Time.

2. Tap Turn On Screen Time, tap Continue, then tap This is My iPad.

3. Tap Downtime, then turn on Downtime.

4. Select Every Day or Customize Days, then set the start and end times.

Set app limits

You can set a time limit for a category of apps (for example, Games or Social Networking) and for individual apps.

1. Go to Settings > Screen Time.

2. If you haven't already turned on Screen Time, tap Turn On Screen Time, tap Continue, then tap This is My iPad.

3. Tap App Limits, then tap Add Limit.

4. Select one or more app categories.

To set limits for individual apps, tap the category name to see all the apps in that category, then select the apps you want to limit. If you select multiple categories or apps, the time limit you set applies to all of them.

5. Tap Next, then set the amount of time allowed.

   To set an amount of time for each day, tap Customize Days, then set limits for specific days.

6. To set a limit for more apps or categories, tap Choose Apps, then repeat step 5.

7. When you finish setting limits, tap Add to return to the App Limits screen.

To temporarily turn off all app limits, tap App Limits on the App Limits screen. To temporarily turn off a limit for a specific category, tap the category, then tap App Limit. To remove a limit for a category, tap the category, then tap Delete Limit.

Set communication limits

You can block incoming and outgoing communication— including phone calls, FaceTime calls, and messages—from specific contacts in iCloud, either at all times or during certain periods.

1. If you haven't already turned on Contacts in iCloud, go to Settings ⚙ > [*your name*] > iCloud, then turn on Contacts.

2. Go to Settings 🌀 > Screen Time.

3. If you haven't already turned on Screen Time, tap Turn On Screen Time, tap Continue, then tap This is My iPad.

4. Tap Communication Limits, then do any of the following:

   - *Limit communication at any time:* Tap During Screen Time, then select Contacts Only, Contacts & Groups with at Least One Contact, or Everyone.

   - *Limit communication during downtime:* Tap During Downtime. The option you selected for During Screen Time is already set here. You can change this setting to Specific Contacts.

     If you select Specific Contacts, tap either Choose From My Contacts or Add New Contact to select people you want to allow communication with during downtime.

If someone who's currently blocked by your Communication Limit settings tries to call you (by phone or FaceTime), or send you a message, their communication won't go through.

If you try to call or send a message to someone who's currently blocked by your Communication Limit settings, their name or number appears in red with a Screen Time hourglass icon, and your communication won't go through. If the limit applies only to downtime, you receive a Time Limit message. You can resume communication with the contact when downtime is over.

To resume communication with contacts who are blocked by your Communication Limit settings, change the settings by following the steps above.

Choose apps you want to allow at all times

You can specify apps that you want to be able to use at any time (for example, in the event of an emergency), even during downtime.

1. Go to Settings > Screen Time.

2. If you haven't already turned on Screen Time, tap Turn On Screen Time, tap Continue, then tap This is My iPad.

3. Tap Always Allowed, then tap or next to an app to add or remove it from the Allowed Apps list.

Set content and privacy restrictions

You can block inappropriate content and set restrictions for iTunes Store and App Store purchases.

1. Go to Settings > Screen Time.

2. If you haven't already turned on Screen Time, tap Turn On Screen Time, tap Continue, then tap This is My iPad.

3. Tap Content & Privacy Restrictions, turn on Content & Privacy Restrictions, then tap options to set content allowances for iTunes Store and App Store purchases, app use, content ratings, and so on.

You can also set a passcode that's required before changing settings.

To share your Screen Time settings and reports across all your devices, make sure you're signed in with the same Apple ID and Share Across Devices is turned on.

Set up Screen Time for a family member on iPad

Screen Time lets you see how family members are using their devices, so you can structure the time they spend on them. You can set up Screen Time for a family member on their device or, if you've set up Family Sharing, you can set up Screen Time for a family member on your device.

Set downtime and app limits on a family member's device

1. On your family member's device, go to Settings > Screen Time.

2. Tap Turn On Screen Time, tap Continue, then tap This is My Child's iPad.

3. To schedule downtime for your family member (time away from the screen), enter the start and end times, then tap Set Downtime.

4. To set limits for categories of apps you want to manage (for example, Games or Social Networking), select the categories.

   To see all the categories, tap Show All Categories.

5. Tap Set, enter an amount of time, then tap Set App Limit.

6. Tap Continue, then enter a Screen Time passcode for managing your family member's Screen Time settings.

Set communication limits on a family member's device

You can block incoming and outgoing communication on your family member's device—including phone calls, FaceTime calls, and messages—from specific contacts, either at all times or during certain periods.

1. If you haven't already turned on Contacts in iCloud on your family member's device, go to Settings > [child's name] > iCloud, then turn on Contacts.

   Note: You can only manage your family member's communication if they're using Contacts in iCloud.

2. On your family member's device, go to Settings > Screen Time.

3. If you haven't already turned on Screen Time, tap Turn On Screen Time, tap Continue, then tap This is My Child's iPad.

4. Tap Communication Limits, then do any of the following:

- *Limit communication at any time:* Tap During Screen Time, then select Contacts Only, Contacts & Groups with at Least One Contact, or Everyone.

- *Limit communication during downtime:* Tap During Downtime. The option you selected for During Screen Time is already set here. You can change this setting to Specific Contacts.

  If you select Specific Contacts, tap either Choose From My Contacts or Add New Contact to select people you want to allow communication with during downtime.

- *Manage a child's contacts:* If you're using Family Sharing, you can manage your child's contacts. Tap Manage [*child's name*] Contacts.

  If your child already has contacts in iCloud, they receive a notification on their device asking them to approve the request to manage them. If they don't have contacts, they don't get a notification and you can immediately add contacts.

When you manage your child's contacts, a new row appears beneath Manage [*child's name*] Contacts to show how many contacts they have. You can view and edit those contacts by tapping that row.

- *Allow contact editing:* Tap Allow Contact Editing to turn off this option and prevent your child from editing their contacts.

  Turning off contact editing and limiting communication at any time to Contacts Only is a good way to control who your child can communicate with and when they can be contacted.

If someone who's currently blocked by the Communication Limit settings tries to call your family member (by phone or FaceTime), or send them a message, their communication won't go through.

If your family member tries to call or send a message to someone who's currently blocked by the Communication Limit settings, the recipient's name or number appears in red with a Screen Time hourglass icon, and the communication won't go through. If the limit applies only to downtime, your family member receives a Time Limit message and can resume communication with the contact when downtime is over.

To allow your family member to communicate with contacts who are blocked by the Communication Limit settings, change the settings by following the steps above.

Choose which apps to allow at all times on a family member's device

You can set which apps you want your family member to be able to use at any time.

1. On your family member's device, go to Settings > Screen Time.

2. If you haven't already turned on Screen Time, tap Turn On Screen Time, tap Continue, then tap This is My Child's iPad.

3. Tap Always Allowed, then tap ⊕ or ⊖ next to an app to add or remove it from the list.

   Note: If your family member needs health or accessibility apps, make sure they're in the Allowed Apps list. If Messages isn't always allowed, your family member may not be able to send or receive messages (including to emergency numbers and contacts) during downtime or after the app limit has expired.

Set content and privacy restrictions on a family member's device

You can help ensure that the content on your family member's device is age appropriate by limiting the explicitness ratings in Content & Privacy Restrictions.

1. On your family member's device, go to Settings > Screen Time.

2. If you haven't already turned on Screen Time, tap Turn On Screen Time, tap Continue, then tap This is My Child's iPad.

3. Tap Content & Privacy Restrictions, then turn on Content & Privacy Restrictions.

4. Choose specific content and privacy options, then tap ⟨.

   Note: To prevent changes to the maximum headphone volume, tap Reduce Loud Sounds, then select Don't Allow.

Add or change Screen Time settings for a family member later

To add or change Screen Time settings later, follow the steps described in Set up Screen Time for yourself on iPad.

Important: If you set up Screen Time for a family member on their device (not through Family Sharing), and you forget the Screen Time passcode, you can use your Apple ID to reset it. However, if you set up Screen Time for a family member on your device through Family Sharing and you forget your Screen Time passcode, you can reset it on your device using your device passcode, Touch ID, or Face ID.

Get a report of your device use on iPad

When you have Screen Time set up, you can get a report of your device use.

1. Go to Settings  > Screen Time.

2. Tap See All Activity, then do any of the following:

- Tap Week to see a summary of your weekly use.

- Tap Day to see a summary of your daily use.

You can also view your summary by tapping a Screen Time Weekly Report notification when it appears on your screen. (If the notification disappears, you can find it in Notification Center. Alternatively, you can add a widget for Screen Time to Today View.)

Chapter 10
Privacy and Security

Sign in with Apple on iPad

With Sign in with Apple, you can sign in to participating apps and websites with your <u>Apple ID</u>. By using your Apple ID to set up and sign in to accounts, you don't need to fill out forms or create and remember new passwords.

Sign in with Apple is designed to respect your privacy. Apps and websites can ask only for your name and email address to set up your account, and Apple won't track you as you use them.

Set up an account and sign in

When a participating app or website asks you to set up an account or to sign in for the first time, do the following:

1. Tap Sign in with Apple.

2. Follow the onscreen instructions.

Some apps (and websites) won't request any personal information from you. In this case, you simply authenticate with Face ID or Touch ID (depending on your model), then start using the app.

Others may ask for your name and email address to set up a personalized account. When an app asks for this information,

Sign in with Apple displays your name and the personal email address from your Apple ID account for you to review.

To edit your name, tap it, then use the keyboard to make changes.

To specify an email address, do one of the following:

- *Use your personal email address:* Tap Share My Email.

  If you have multiple email addresses associated with your Apple ID, choose the address you want.

- *Hide your email address:* Tap Hide My Email.

  When you choose this option, Apple creates a unique, anonymized address for you that forwards email from the app to your personal address. This option allows you to receive email from the app without sharing your personal email address.

After you review your information and choose an email option, tap Continue, authenticate with Face ID or Touch ID (depending on your model), then start using the app.

Sign in to access your account

After you set up an account with an app or website using Sign in with Apple, you typically don't need to sign in to it again on your iPad. But if you're asked to sign in (for example, after you sign out of an account), do the following:

1. Tap Sign in with Apple.

2. Review the Apple ID that appears, then tap Continue.

3. Authenticate with Face ID or Touch ID (depending on your model).

Change the address used to forward email

If you chose to hide your email address when you created an account and you have more than one address associated with your Apple ID, you can change the address that receives your forwarded email.

1. Go to Settings > [*your name*] > Name, Phone Numbers, Email > Forward To.

2. Choose a different email address, then tap Done.

Change Sign in with Apple settings for an app or website

1. Go to Settings > [*your name*] > Password and Security.

2. Tap Apps Using Your Apple ID.

3. Choose an app, then do either of the following:

- *Turn off forwarding email:* Turn off Forward To. You won't receive any further emails from the app.

- *Stop using Sign in with Apple:* Tap Stop Using Apple ID. You may be asked to create a new account the next time you try to sign in with the app.

Sign in with Apple also works on your other devices—iPhone, Apple Watch, Mac, Apple TV, and iPod touch—where you're signed in with the same Apple ID.

To sign in from an Android app, a Windows app, or any web browser, tap Sign in with Apple, then enter your Apple ID and password.

Sign in with Apple requires two-factor authentication for your Apple ID. This protects your Apple ID, your app accounts, and your app content.

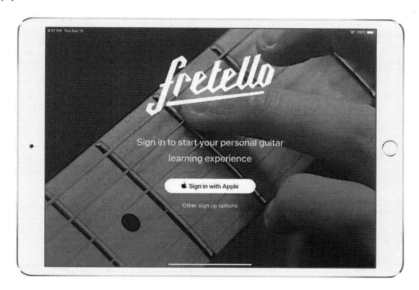

Set which apps can access your location on iPad

With Location Services, you can choose which location-based apps—for example, Reminders, Maps, and Camera—can gather and use data indicating your location. Your approximate location is determined using information about your local Wi-Fi networks (if you have Wi-Fi turned on), your cellular network (WiFi + Cellular models, if you have cellular data turned on), and GPS

(WiFi + Cellular models, if available). When an app is using Location Services, ⊲ appears in the status bar.

Turn on Location Services

- If you didn't turn on Location Services when you first set up iPad, go to Settings ⚙ > Privacy > Location Services, then turn on Location Services.

Turn off Location Services

- Go to Settings ⚙ > Privacy > Location Services, then choose from the options to turn off Location Services for some apps and services, or for all of them.

If you turn off Location Services, you're asked to turn it on again the next time an app or service tries to use it. Review the terms and privacy policy for each third-party app to understand how it uses the data it's requesting.

Hide the map in Location Services alerts

When you allow an app to always use your location in the background, you may receive alerts about the app's use of that information. (These alerts let you change your permission, if you want to.) In the alerts, a map shows locations recently accessed by the app.

To hide the map, go to Settings ⚙ > Privacy > Location Services > Location Alerts, then turn off Show Map in Location Alerts.

With the setting off, you continue to receive location alerts, but the map isn't shown.

Change Location Services settings for system services

Several system services, such as location-based suggestions and location-based ads, use Location Services.

To see the status for each service, to turn Location Services on or off for each service, or to show ⌲ in the status bar when enabled system services use your location, go to Settings ⚙ > Privacy > Location Services > System Services.

Delete significant locations

The Maps app keeps track of the places you've recently visited, as well as when and how often you visited them. Maps uses this information to provide you with personalized services like predictive traffic routing. You can delete this information.

1. Go to Settings ⚙ > Privacy > Location Services > System Services > Significant Locations.

2. Do one of the following:

   - *Delete a single location:* Tap the location, tap Edit, then tap ⊖.

   - *Delete all locations:* Tap Clear History. This action clears all your significant locations on any devices that are signed in with the same Apple ID.

Change app access to private data on iPad

Change which apps and features have access to private information in Contacts, Calendar, Reminders, Photos, Home, and more.

1. Go to Settings 🎛 > Privacy.

2. Tap a category of data, such as Contacts, Photos, Microphone, or Motion & Fitness.

   The list shows the apps and features that requested access to that data, along with the level of access that you allowed.

3. Tap an app or feature, then choose a different level of access or deny access.

Review the terms and privacy policy for each third-party app to understand how it uses the data it's requesting.

Limit ad targeting on iPad

Turn off location-based ads, reset or limit targeted advertising in App Store and News, and prevent cross-site tracking in Safari.

Turn off location-based ads and offers

- Go to Settings 🎛 > Privacy > Location Services > System Services, then turn off Location-Based Apple Ads.

Reset or limit ad tracking

Go to Settings 🔘 > Privacy > Advertising (at the bottom of the screen), then do any of the following:

- *Clear the data used to determine which ads might be relevant to you:* Tap Reset Advertising Identifier.

- *Opt out of targeted advertising:* Turn on Limit Ad Tracking.

   Note: When you turn on Limit Ad Tracking, you may still receive the same number of ads, but they may be less relevant to you.

View the information Apple uses to deliver targeted ads

Go to Settings 🔘 > Privacy > Advertising > View Ad information.

The information is used by Apple to deliver more relevant ads in the App Store and News. Your personal data isn't provided to other parties.

Keep your Safari browsing activities to yourself

While browsing the web, you can prevent cross-site tracking, block cookies, erase your browsing history, and more.

Go to Settings 🔘 > Privacy > Advertising > About Advertising & Privacy.

Set a passcode on iPad

For better security, set a passcode that needs to be entered to unlock iPad when you turn it on or wake it. Setting a passcode turns on data protection, which encrypts your iPad data with

256-bit AES encryption. (Some apps may opt out of using data protection.)

Set or change the passcode

1. Go to Settings ⚙, then depending on your model, tap one of the following:

   • Face ID & Passcode

   • Touch ID & Passcode

   • Passcode

   Tap Turn Passcode On or Change Passcode.

To view options for creating a password, tap Passcode Options. The most secure options are Custom Alphanumeric Code and Custom Numeric Code.

After you set a passcode, on supported models you can use Face ID or Touch ID to unlock iPad. For additional security, however, you must always enter your passcode to unlock your iPad under the following conditions:

• You turn on or restart your iPad.

• You haven't unlocked your iPad for more than 48 hours.

• You haven't unlocked your iPad with the passcode in the last 6.5 days, and you haven't unlocked it with Face ID or Touch ID in the last 4 hours.

• Your iPad receives a remote lock command.

- There are five unsuccessful attempts to unlock your iPad with Face ID or Touch ID.

Change when iPad automatically locks

- Go to Settings 🎛 > Display & Brightness > Auto-Lock, then set a length of time.

Erase data after 10 failed passcodes

Set iPad to erase all information, media, and personal settings after 10 consecutive failed passcode attempts.

1. Go to Settings 🎛, then depending on your model, tap one of the following:

   - Face ID & Passcode

   - Touch ID & Passcode

   - Passcode

   Turn on Erase Data.

After all data is erased, you must restore iPad from a backup or set it up again as new.

Turn off the passcode

1. Go to Settings 🎛, then depending on your model, tap one of the following:

   - Face ID & Passcode

   - Touch ID & Passcode

   - Passcode

Tap Turn Passcode Off.

Reset the passcode

If you enter the wrong passcode six times in a row, you'll be locked out of your device, and you'll receive a message that says iPad is disabled. If you can't remember your passcode, you can erase your iPad with a computer or with recovery mode, then set a new passcode. (If you made an iCloud or computer backup before you forgot your passcode, you can restore your data and settings from the backup.)

Set up Face ID on iPad

Use Face ID (<u>supported models</u>) to unlock iPad, authorize purchases and payments, and sign in to many third-party apps by simply glancing at your iPad.

To use Face ID, you must also <u>set a passcode</u> on your iPad.

Set up Face ID or add an alternate appearance

- If you didn't set up Face ID when you first set up your iPad, go to Settings 🔘 > Face ID & Passcode > Set up Face ID, then follow the onscreen instructions.

- To set up an additional appearance for Face ID to recognize, go to Settings > Face ID & Passcode > Set Up an Alternate Appearance, then follow the onscreen instructions.

If you have physical limitations, you can tap Accessibility Options during Face ID set up. When you do this, setting up facial recognition doesn't require the full range of head motion. Using Face ID is still secure, but it requires more consistency in how you look at iPad.

Face ID also has an accessibility feature you can use if you're blind or have low vision. If you don't want Face ID to require that you look at iPad with your eyes open, go to Settings > Accessibility > Face ID & Attention, then turn off Require Attention for Face ID. This feature is automatically turned off if you turn on VoiceOver when you first set up iPad.

Temporarily disable Face ID

You can temporarily prevent Face ID from unlocking your iPad.

1. Press and hold the top button and either volume button for 2 seconds.

2. After the sliders appear, press the top button to immediately lock iPad.

iPad locks automatically if you don't touch the screen for a minute or so. The next time you unlock iPad with your passcode, Face ID is enabled again.

Turn off Face ID

1. Go to Settings 🎛 > Face ID & Passcode.

2. Do one of the following:

   - *Turn off Face ID for specific items only:* Turn off one or more options: iPad Unlock, Apple Pay, iTunes & App Store, or Safari AutoFill.

   - *Turn off Face ID:* Tap Reset Face ID.

If your device is lost or stolen, you can prevent Face ID from being used to unlock your device with **Find My iPhone** Lost Mode.

Set up Touch ID on iPad

Use Touch ID (<u>supported models</u>) to unlock iPad, authorize purchases and payments, and sign in to many third-party apps by pressing the Home button with your finger or thumb. To use Touch ID, you must <u>set a passcode</u> on your iPad.

Turn on fingerprint recognition

1. If you didn't turn on fingerprint recognition when you first set up your iPad, go to Settings 🎛 > Touch ID & Passcode.

2. Turn on any of the options, then follow the onscreen instructions.

If you turn on iTunes & App Store, you're asked for your <u>Apple ID</u> password when you make your first purchase from the iTunes Store, the App Store, or Apple Books. When you make your next purchases, you're asked to use Touch ID.

Add a fingerprint

You can add multiple fingerprints (both of your thumbs and forefingers, for example).

1. Go to Settings > Touch ID & Passcode.

2. Tap Add a Fingerprint.

3. Follow the onscreen instructions.

Name or delete a fingerprint

1. Go to Settings > Touch ID & Passcode.

   If you added more than one fingerprint, place a finger on the Home button to identify its print.

2. Tap the fingerprint, then enter a name (such as "Thumb") or tap Delete Fingerprint.

Unlock iPad by touching instead of pressing the Home button

Go to Settings > Accessibility > Home Button, then turn on Rest Finger to Open.

Turn off Touch ID

Go to Settings 🔘 > Touch ID & Passcode, then turn off one or more of the options.

Chapter 11
Restart, Update, Reset and Restore

Restart iPad

If your iPad isn't working right, try restarting it.

Turn iPad off and on

1. To turn off iPad, do one of the following:

- *Models with the Home button:* Press and hold the top button until the slider appears, then drag the slider.

- *Other models:* Simultaneously press and hold the top button and either volume button until the sliders appear, then drag the top slider.

- *All models:* Go to Settings ⊚ > General > Shut Down, then drag the slider.

  To turn iPad back on, press and hold the top button until the Apple logo appears.

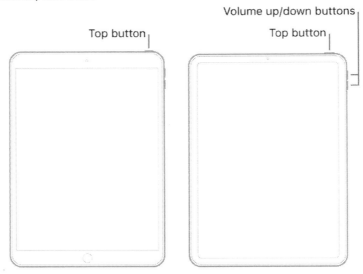

Force restart iPad

If iPad isn't responding, do one of the following:

- *Models with the Home button:* Press and hold the top button and the Home button at the same time. When the Apple logo appears, release both buttons.

- *Other models:* Press and release the volume up button, press and release the volume down button, then press and hold the top button. When the Apple logo appears, release the button.

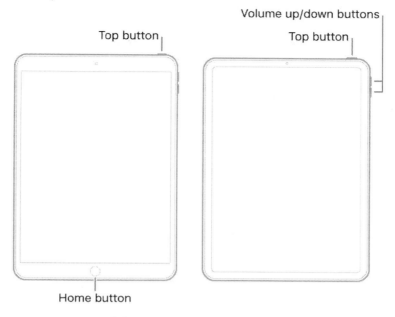

Update iPadOS

When you update to the latest version of iPadOS, your data and settings remain unchanged.

Note: Before you update, set up iPad to <u>back up</u> automatically, or back up your iPad manually.

Update iPad automatically

If you didn't turn on automatic updates when you first set up iPad, go to Settings ⚙ > General > Software Update > Automatic Updates, then turn on Automatic Updates.

iPad automatically installs updates wirelessly, and you're notified before the update occurs. To see the currently installed version of iPadOS, go to Settings > General > Software Update.

Update iPad manually

Go to Settings ⚙ > General > Software Update > Automatic Updates, then turn off Automatic Updates.

At any time, you can check for and install software updates. Go to Settings > General > Software Update. The screen shows the currently installed version of iPadOS and whether an update is available.

Update using your computer

1. Connect iPad and your computer using USB.

2. Do one of the following:

   - *In the Finder sidebar on your Mac:* Select your iPad, then click General at the top of the window.

     Note: To use the Finder to update your iPad, macOS Catalina is required. With earlier versions of macOS, use iTunes to update your iPad.

   - *In the iTunes app on a Windows PC:* Click the iPad button near the top left of the iTunes window, then click **Summary**.

   Click Check for Update.

   To install an available update, click Update.

Back up iPad

You can back up iPad using iCloud or your computer.

Tip: If you replace your iPad, you can use its backup to transfer your information to the new device.

Back up iPad using iCloud

1. Go to Settings ⚙ > [*your name*] > iCloud > iCloud Backup.

2. Turn on iCloud Backup. iCloud automatically backs up your iPad daily when iPad is connected to power, locked, and on Wi-Fi.

3. To perform a manual backup, tap Back Up Now.

To view your iCloud backups, go to Settings > [*your name*] > iCloud > Manage Storage > Backups. To delete a backup, choose a backup from the list, then tap Delete Backup.

Back up iPad using your Mac

1. Connect iPad and your computer using USB.

2. In the Finder sidebar on your Mac, select your iPad.

   Note: To use the Finder to back up iPad, macOS Catalina is required. With earlier versions of macOS, use iTunes to back up iPad.

3. At the top of the Finder window, click General.

4. Select "Back up all of the data on your iPad to this Mac."

5. To encrypt your backup data and protect it with a password, select "Encrypt local backup."

6. Click Back Up Now.

Note: You can also connect iPad to your computer wirelessly if you set up Wi-Fi syncing.

Back up iPad using your Windows PC

1. Connect iPad and your computer using USB.

2. In the iTunes app on your PC, click the iPad button near the top left of the iTunes window.

3. Click Summary.

4. Click Back Up Now (below Backups).

5. To encrypt your backups, select "Encrypt local backup," type a password, then click Set Password.

To see the backups stored on your computer, choose Edit > Preferences, then click Devices. Encrypted backups have a lock icon in the list of backups.

Return iPad settings to their defaults

You can return settings to their defaults without erasing your content.

If you want to save your settings, **back up iPad** before returning them to their defaults. For example, if you're trying to solve a problem but returning settings to their defaults doesn't help, you might want to restore your previous settings from a backup.

1. Go to Settings 🔘 > General > Reset.

2. Choose an option:

   WARNING: If you choose the Erase All Content and Settings option, all of your content is removed.

Reset All Settings: All settings—including network settings, the keyboard dictionary, the Home screen layout, location settings, privacy settings, and Apple Pay cards—are removed or reset to their defaults. No data or media are deleted.

- **Reset Network Settings**: Only network settings are removed.

  When you reset network settings, previously used networks and VPN settings that weren't installed by a configuration profile or mobile device management (MDM) are removed. Wi-Fi is turned off and then back on, disconnecting you from any network you're on. The Wi-Fi and Ask to Join Networks settings remain turned on.

  To remove VPN settings installed by a configuration profile, go to Settings > General > Profiles & Device Management, select the configuration profile, then tap Remove Profile. This also removes other settings and accounts provided by the profile.

  To remove network settings installed by MDM, go to Settings > General > Profiles & Device Management, select the management, then tap Remove Management. This also removes other settings and certificates provided by MDM.

- **_Reset Keyboard Dictionary_**_:_ You add words to the keyboard dictionary by rejecting words iPad suggests as you type. Resetting the keyboard dictionary erases only the words you've added.

  - **_Reset Home Screen Layout_**_:_ Returns the built-in apps to their original layout on the Home screen.

  - **_Reset Location & Privacy_**_:_ Resets the location services and privacy settings to their defaults.

If you want to use a computer to delete your content along with your settings and then restore iPad to factory settings.

Restore all content to iPad from a backup

You can restore content, settings, and apps from a backup to a new or newly erased iPad.

Important: You must first create a backup of your iPad.

Restore iPad from an iCloud backup

1. Turn on a new or newly erased iPad.

2. Follow the online instructions to choose a language and region.

3. Tap Set Up Manually.

4. Tap Restore from iCloud Backup, then follow the onscreen instructions.

You're asked for your Apple ID.

Restore iPad from a computer backup

1. Using USB, connect a new or newly erased iPad to the computer containing your backup.

2. Do one of the following:

   • *In the Finder sidebar on your Mac:* Select your iPad, then click Trust.

     Note: To use the Finder to restore iPad from a backup, macOS Catalina is required. With earlier versions of macOS, **use iTunes** to restore from a backup.

   • *In the iTunes app on a Windows PC:* If you have multiple devices connected to your PC, click the device icon near the top left of the iTunes window, then select your new or newly erased iPad from the list.

   On the welcome screen, click "Restore from this backup," choose your backup from the list, then click Continue.

If your backup is encrypted, you must enter the password before restoring your files and settings.

Restore purchased and deleted items to iPad

You can redownload items from the App Store, Book Store, and iTunes Store without repurchasing them. If you're part of a **Family Sharing** group, you can download items purchased by other family members, too. **You** can also recover recently deleted email, photos, notes, and voice memos.

Sell or give away your iPad

Before you sell or give away your iPad, be sure to perform the following tasks:

- **Back up iPad**. If you replace one iPad with another, you can use the setup assistant to restore the backup to your new iPad.

- **Erase all content and settings** that contain personal information. If you previously **turned on Find My** for your iPad, Activation Lock is removed when you erase iPad, making it ready for a new owner.

Erase all content and settings from iPad

When you delete data, it's no longer accessible through the iPad interface, but it isn't erased from iPad storage. To remove all of your content and settings from storage, erase iPad. For example, erase iPad before you **sell it or give it away**. If you want to save your content and settings, **back up iPad** before erasing it.

1. Go to Settings ⚙ > General > Reset.

2. Tap Erase All Content and Settings.

When iPad restarts with all content and settings erased, you have the option to set up iPad as new or restore it from a backup.

Restore iPad to factory settings

You can use a Mac or Windows PC to erase all data and settings from your iPad, restore iPad to factory settings, and install the latest version of iPadOS.

Important: Restoring your iPad to its factory settings erases all your data and settings. However, before iPad is erased, you have the option to back up your iPad. If you make a backup, you can use the backup to restore your data and settings on your iPad or on a new device.

1. Connect iPad and your computer using USB.

2. Do one of the following:

   - *In the Finder sidebar on your Mac:* Select your iPad, click General at the top of the window, then click Restore iPad.

     Note: To use the Finder to restore iPad to factory settings, macOS Catalina is required. With earlier versions of macOS, use iTunes to restore iPad.

   - *In the iTunes app on a Windows PC:* Click the iPad button near the top left of the iTunes window, click Summary, then click Restore iPad.

     Follow the onscreen instructions.

Install or remove configuration profiles on iPad

Configuration profiles define settings for using iPad with corporate or school networks or accounts. You might be asked to install a configuration profile that was sent to you in an email, or one that is downloaded from a webpage. You're asked for permission to install the profile and, when you open the file, information about what it contains is displayed. You can see the profiles you have installed in Settings > General > Profiles & Device Management. If you delete a profile, all of the settings, apps, and data associated with the profile are also deleted.

Chapter 12

How to use the Accessibility features

Get started with accessibility features on iPad.

iPad provides many accessibility features to support your vision, physical and motor, hearing, and learning needs. Learn how to configure these features and set up shortcuts for easy access.

Turn on accessibility features during setup

You can turn on many accessibility features right away when you first set up iPad. Turn on iPad, then do any of the following:

- *Turn on VoiceOver:* Triple-click the top button (iPad Pro (11-inch) and iPad Pro (12.9-inch) (3rd generation)) or triple-click the Home button (other models).

- *Turn on Zoom:* Double-tap the screen with three fingers.

- *Turn on Switch Control, Larger Text, Smart Invert, and more:* Choose a language and country, tap ⊕, then choose the features you want.

Change accessibility settings

After you set up iPad, you can adjust accessibility settings.

1. Go to Settings 🌐 > Accessibility.

2. Choose any of the following features:

- Vision

    - VoiceOver

    - Zoom

    - Magnifier

    - Display & Text Size

    - Motion

    - Spoken Content

    - Audio Descriptions

- Physical and Motor

    - Touch

    - Face ID & Attention

    - Switch Control

    - Voice Control

    - Home or top button

    - Apple TV Remote

    - Keyboards

    - Apple Pencil

- Hearing

- Hearing Devices

- RTT

- Audio/Visual

- Subtitles & Captioning

- General

  - Guided Access

  - Siri

  - Accessibility Shortcut

Turn on and practice VoiceOver on iPad

With VoiceOver—a gesture-based screen reader—you can use iPad even if you don't see the screen. VoiceOver gives audible descriptions of what's on your screen—from battery level, to who's calling, to which app your finger is on. You can also adjust the speaking rate and pitch to suit your needs.

When you touch the screen or drag your finger over it, VoiceOver speaks the name of the item your finger is on, including icons and text. To interact with the item, such as a button or link, or to navigate to another item, use VoiceOver gestures.

When you go to a new screen, VoiceOver plays a sound, then selects and speaks the name of the first item on the screen (typically in the top-left corner). VoiceOver tells you when the display changes to landscape or portrait orientation, when the screen becomes dimmed or locked, and what's active on the Lock screen when you wake iPad.

Turn VoiceOver on or off

Important: VoiceOver changes the gestures you use to control iPad. When VoiceOver is on, you must use VoiceOver gestures to operate iPad.

To turn VoiceOver on or off, use any of the following methods:

- Go to Settings ⚙ > Accessibility > VoiceOver, then turn the setting on or off.

- **Summon Siri** and say "Turn on VoiceOver" or "Turn off VoiceOver."

- **Triple-click the Home button** (models with the Home button).

- **Triple-click the top button** (other models).

- Use Control Center.

Learn and practice VoiceOver gestures

You can practice VoiceOver gestures in a special area without affecting iPad or its settings. When you practice a gesture, VoiceOver describes the gesture and the resulting action.

Try different techniques to discover which works best for you. If a gesture doesn't work, try a quicker movement, especially for a double-tap or swipe gesture. To swipe, try brushing the screen quickly with your finger or fingers. For best results using multifinger gestures, touch the screen with some space between your fingers.

1. Go to Settings ⚙ > Accessibility > VoiceOver.

2. Turn on VoiceOver, tap VoiceOver Practice, then double-tap to start.

3. Practice the following gestures with one, two, three, and four fingers:

   • Tap

   • Double-tap

   • Triple-tap

   • Swipe left, right, up, or down

   When you finish practicing, tap Done, then double-tap to exit.

Change your VoiceOver settings on iPad

You can customize the settings for VoiceOver, such as the audio options, language, voice, speaking rate, and verbosity.

Adjust the VoiceOver volume and other audio options

• To increase or decrease the volume, press the volume buttons on iPad.

• To set other audio options, go to Settings ⚙ > Accessibility > VoiceOver > Audio, then do any of the following:

- Turn on Mute Sound Effects.

- Turn on Audio Ducking to temporarily reduce playback volume when VoiceOver speaks.

- Adjust audio routing options when you connect additional devices, such as an instrument amplifier or a DJ mixer.

Set the VoiceOver language

VoiceOver uses the same language you choose for your iPad. VoiceOver pronunciation of some languages is affected by the Region Format you choose.

1. Go to Settings ⚙ > General > Language & Region.

2. Tap iPad Language, then choose a language.

Adjust the speaking voice

Go to Settings ⚙ > Accessibility > VoiceOver, then do any of the following:

- *Adjust the speaking rate:* Drag the Speaking Rate slider.

- *Choose a voice:* Tap Speech > Voice, then choose a voice. To download an enhanced voice, tap ⬇.

- *Adjust the pitch:* Tap Speech, then drag the slider. You can also turn on Use Pitch Change to have VoiceOver use a higher pitch when speaking the first item of a group (such as a list or table) and a lower pitch when speaking the last item of a group.

- *Specify the pronunciation of certain words:* Tap Speech > Pronunciations, tap ✛, enter a phrase, then dictate or spell out how you want the phrase to be pronounced.

  Note*:* You can dictate only if you turned on Enable Dictation in Settings > General > Keyboards.

Set how much VoiceOver tells you

Go to Settings ⚙ > Accessibility > VoiceOver, then tap any of the following:

- *Verbosity:* Choose options to have VoiceOver speak hints, punctuation, uppercase letters, embedded links, and more. VoiceOver can even confirm rotor actions.

  To change how VoiceOver speaks punctuation, tap Punctuation, then choose a group. You can also create new groups—for example, a programming group in which "[" is spoken as "left brack."

- *Always Speak Notifications:* VoiceOver reads notifications, including incoming text messages as they occur, even if iPad is locked. Unacknowledged notifications are repeated when you unlock iPad.

Customize VoiceOver settings for an activity

You can customize a group of VoiceOver settings for an activity such as programming. Apply the settings automatically when you open certain apps or by adjusting the rotor.

1. Go to Settings ⦿ > Accessibility > VoiceOver > Activities.

2. Choose an existing activity or tap Add Activity.

3. Adjust settings for speech, verbosity, and braille.

4. Choose Apps or Context to automatically apply the settings for this activity.

Adjust VoiceOver visuals

Go to Settings ⦿ > Accessibility > VoiceOver, then turn on any of the following:

- *Large Cursor:* If you have trouble seeing the black outline around the selected item, you can enlarge and thicken the outline.

- *Caption Panel:* The text spoken by VoiceOver is displayed at the bottom of the screen.

Learn VoiceOver gestures on iPad

When VoiceOver is on, standard touchscreen gestures have different effects, and additional gestures let you move around the screen and control individual items. VoiceOver gestures include two-, three-, and four-finger taps and swipes.

You can use different techniques to perform VoiceOver gestures. For example, you can perform a two-finger tap using two fingers on one hand, one finger on each hand, or your thumbs. Instead of selecting an item and double-tapping, you can use a split-tap gesture—touch and hold an item with one finger, then tap the screen with another finger.

Explore and speak items on the screen

To explore the screen, drag your finger over it. VoiceOver speaks the name of each item you touch.

You can also use VoiceOver gestures to explore the screen in order, from top to bottom and left to right.

Operate iPad using VoiceOver gestures

When VoiceOver is on, you need to use special gestures to unlock iPad, go to the Home screen, open Control Center, switch apps, and more.

Unlock iPad

- **Models with Face ID**: Wake iPad and glance at it, then drag up from the bottom edge of the screen until you hear two rising tones.

- **Models with Touch ID**: Press the Home button.

- **Other models**: Press the Home button, then enter your passcode.

To avoid having your passcode spoken as you enter it, enter your passcode silently using handwriting mode or type onscreen braille.

Go to the Home screen

- Drag one finger up from the bottom edge of the screen until you hear two rising tones, then lift your finger.

- Press the Home button (models with the Home button).

Use the dock

- Slide one finger up from the bottom edge until you hear two rising tones, then swipe down.

- Switch to another app

- Swipe right or left with five fingers to cycle through the open apps. (Make sure Gestures is turned on in Settings ⚙ > General > Multitasking & Dock.)

Alternatively, you can use the App Switcher:

1. Open the App Switcher using one of the following methods:

   - Drag one finger up from the bottom edge of the screen until you hear three tones, then lift your finger.

   - Double-click the Home button (models with the Home button).

   To browse the open apps, swipe left or right until the app you want is selected.

   Double-tap to open the app.

Open Control Center

- Drag one finger down from the top edge of the screen until you hear two rising tones.

- Tap any item in the status bar, then swipe up with three fingers.

To dismiss Control Center, do a two-finger scrub.

View notifications

- Drag one finger down from the top edge of the screen until you hear three rising tones.

- Tap any item in the status bar, then swipe down with three fingers.

To dismiss the notifications screen, do a two-finger scrub.

Speak status bar information

1. Tap the status bar at the top of the screen.

2. Swipe left or right to hear the time, battery state, Wi-Fi signal strength, and more.

Rearrange apps on your Home screen

Use one of the following methods:

- *Drag and drop:* Tap an icon on the Home screen, then double-tap and hold your finger on the screen until you hear three rising tones. The item's relative location is described as you drag. Lift your finger when the icon is in its new location. Drag an icon to the edge of the screen to move it to another Home screen.

- *Move actions:* Tap an app, then swipe down to hear available actions. When you hear "Edit Mode," double-tap to start arranging apps. Find the app you want to move, then swipe down to the Move action and double-tap. Move the VoiceOver cursor to the new destination for the app, then choose from the available actions: Cancel Move, Create New Folder, Add to Folder, Move Before, or Move After. 0

When you're finished, tap Done, then double-tap.

Search from the Home screen

1. Tap anywhere on the Home screen outside the status bar.

2. Swipe down with three fingers.

Control VoiceOver using the rotor on iPad

You can use the VoiceOver rotor to change how VoiceOver works. You can adjust the VoiceOver volume or speaking rate, move from one item to the next on the screen, select special input methods such as Braille Screen Input or Handwriting, and more.

When you use Magic Keyboard to control VoiceOver, use the rotor to adjust settings such as volume, speech rate, use of pitch or phonetics, typing echo, and reading of punctuation.

Use the VoiceOver rotor

1. When VoiceOver is turned on, rotate two fingers on your screen as if you're turning a dial. If you prefer to use one finger on each hand, simultaneously drag up with one finger and drag down with the other.

   VoiceOver speaks the rotor setting. Keep rotating your fingers to hear more settings. Stop rotating your fingers when you hear the setting you want.

2. Swipe your finger up or down on the screen to use the setting.

   The available rotor settings and their effects depend on what you're doing. For example, if you choose Headings when you're browsing a webpage, a swipe down or up will move the VoiceOver cursor to the next or previous heading.

Customize the VoiceOver rotor

1. Go to Settings ⚙ > Accessibility > VoiceOver.

2. Do any of the following:

   - *Add or reorder the rotor settings:* Tap Rotor, then choose the settings you want, or drag ≡ to reorder settings.

   - *Add another language:* Tap Speech > Add New Language (below Rotor Languages), then choose a language.

   - *Have VoiceOver confirm rotor actions:* Tap Verbosity, then turn on Speak Confirmation.

Use the onscreen keyboard with VoiceOver on iPad

VoiceOver changes how you use the onscreen keyboard when you activate an editable text field. You can enter, select, and delete text; change the keyboard language; and more.

Enter text with the onscreen keyboard

1. Select a text field, then double-tap.

   The insertion point and the onscreen keyboard appear.

2. Enter text using one of the following methods:

   - *Standard typing (default):* Select a key on the keyboard by swiping left or right, then double-tap to enter the character. Or move your finger around the keyboard to select a key and, while continuing to touch the key with one finger, tap the screen with another finger. VoiceOver speaks the key when it's selected, and again when the character is entered.

   - *Touch typing:* Touch a key on the keyboard to select it, then lift your finger to enter the character. If you touch the wrong key, slide your finger to the key you want. VoiceOver speaks the character for each key as you touch it, but doesn't enter a character until you lift your finger.

   - *Direct Touch typing:* VoiceOver is disabled for the keyboard only, so you can type just as you do when VoiceOver is off.

- *Dictation:* Use a two-finger double-tap on the keyboard to start and stop dictation.

To enter an accented character, use one of the following methods:

- *Standard typing (default):* Select the plain character, then double-tap and hold until you hear a tone indicating alternate characters have appeared. Drag left or right to select and hear the choices. Release your finger to enter the current selection.

- *Touch typing:* Touch and hold a character until the alternate characters appear.

Edit text with the onscreen keyboard

- *Move the insertion point:* Swipe up or down to move the insertion point forward or backward in the text. Use the rotor to choose whether you want to move the insertion point by character, by word, or by line. To jump to the beginning or end, double-tap the text.

VoiceOver makes a sound when the insertion point moves, and speaks the character, word, or line that the insertion point moves across. When moving forward by words, the insertion point is placed at the end of each word, before the space or punctuation that follows. When moving backward, the insertion point is placed at the end of the preceding word, before the space or punctuation that follows it.

- *Move the insertion point past the punctuation at the end of a word or sentence:* Use the rotor to switch back to character mode.

  When moving the insertion point by line, VoiceOver speaks each line as you move across it. When moving forward, the insertion point is placed at the beginning of the next line (except when you reach the last line of a paragraph, when the insertion point is moved to the end of the line just spoken). When moving backward, the insertion point is placed at the beginning of the line that's spoken.

- *Delete a character:* Use ⌫ .

- *Select text:* Use one of the following methods.

  - Set the rotor to Text Selection, swipe up or down to choose Character, Word, Line, or Sentence, then swipe left or right to move backward or forward. (You may need to enable Text Selection—go to Settings ⚙ > Accessibility > VoiceOver > Rotor.)

  - Set the rotor to Edit, swipe up or down to choose Select or Select All, then double-tap. If you choose Select, the word closest to the insertion point is selected when you double-tap. To increase or decrease the selection, do a two-finger scrub to dismiss the pop-up menu, then pinch.

Cut, copy, or paste: Set the rotor to Edit, select the text, swipe up or down to choose Cut, Copy, or Paste, then double-tap.

Fix misspelled words: Set the rotor to Misspelled Words, then swipe up or down to jump to the previous or next misspelled word. Swipe left or right to choose a suggested replacement, then double-tap to use the replacement.

Undo: Shake iPad, swipe left or right to choose the action to undo, then double-tap.

Change the keyboard settings

1. Go to Settings ⚙ > Accessibility > VoiceOver.

2. Tap any of the following:

   • *Typing Style:* You can choose a new style. Or, set the rotor to Typing Mode, then swipe up or down.

   • *Phonetic Feedback:* Speak text character by character. VoiceOver first speaks the character, then its phonetic equivalent—for example, "f" and then "foxtrot.

   • *Typing Feedback:* Choose to speak characters, words, both, or nothing.

   • *Rotor:* Select the settings you want to include in the rotor.

   • *Speech:* Tap Add New Language (below Rotor Languages), then choose a language.

- *Verbosity:* Tap Deleting Text. To have VoiceOver speak deleted characters in a lower pitch, tap Change Pitch.

Chapter 13

Safety, Warning and Support

Important safety information for iPad

WARNING: Failure to follow these safety instructions could result in fire, electric shock, injury, or damage to iPad or other property. Read all the safety information below before using iPad.

Handling - Handle iPad with care. It is made of metal, glass, and plastic and has sensitive electronic components inside. iPad or its battery can be damaged if dropped, burned, punctured, or crushed, or if it comes in contact with liquid. If you suspect damage to iPad or the battery, discontinue use of iPad, as it may cause overheating or injury. Don't use iPad with a cracked screen, as it may cause injury. If you're concerned about scratching the surface of iPad, consider using a case or cover.

Repairing - Don't open iPad and don't attempt to repair iPad yourself. Disassembling iPad may damage it or may cause injury to you. iPad Pro (11-inch) and iPad Pro (12.9-inch) (3rd generation) contain lasers that could be damaged during repair or disassembly, which could result in hazardous exposure to infrared laser emissions that are not visible. If iPad is damaged, malfunctions, or comes in contact with liquid, contact Apple or an Apple Authorized Service Provider. Repairs by service providers other than Apple or an Apple Authorized Service Provider may not involve the use of Apple genuine parts and may affect the safety and functionality of the device.

Battery - Don't attempt to replace the iPad battery yourself. The lithium-ion battery in iPad should be replaced by Apple or an authorized service provider. Improper replacement or repair could damage the battery, cause overheating, or result in injury. The battery must be recycled or disposed of separately from household waste. Don't incinerate the battery.

Distraction - Using iPad in some circumstances may distract you and might cause a dangerous situation (for example, avoid using headphones while riding a bicycle and avoid typing a text message while driving a car). Observe rules that prohibit or restrict the use of mobile devices or headphones.

Navigation - Maps depends on data services. These data services are subject to change and may not be available in all regions, resulting in maps and location-based information that may be unavailable, inaccurate, or incomplete. Compare the information provided in Maps to your surroundings. Use common sense when navigating. Always observe current road conditions and posted signs to resolve any discrepancies. Some Maps features require Location Services.

Charging - Charge iPad with the included USB cable and power adapter. You can also charge iPad with "Made for iPad" or other third-party cables and power adapters that are compliant with

USB 2.0 or later and with applicable country regulations and international and regional safety standards, including the International Standard for Safety Information Technology Equipment (IEC 60950-1). Other adapters may not meet applicable safety standards, and charging with such adapters could pose a risk of death or injury.

Using damaged cables or chargers, or charging when moisture is present, can cause fire, electric shock, injury, or damage to iPad or other property. When you use the Apple USB power adapter to charge iPad, make sure the USB cable is fully inserted into the power adapter before you plug the adapter into a power outlet. It's important to keep iPad and its power adapter in a well-ventilated area when in use or charging.

Charging cable and connector - Avoid prolonged skin contact with the charging cable and connector when the charging cable is connected to a power source because it may cause discomfrt or injury. Sleeping or sitting on the charging cable or connector should be avoided.

Prolonged heat exposure - iPad and its USB power adapter comply with applicable surface temperature standards and limits defined by the International Standard for Safety of Information Technology Equipment (IEC 60950-1). However, even within these limits, sustained contact with warm surfaces for long periods of time may cause discomfort or injury. Use common sense to avoid situations where your skin is in contact with a device or its power adapter when it's operating or connected to a power source for long periods of time. For example, don't sleep on a device or power adapter, or place them under a blanket, pillow, or your body, when it's connected to a power source. Keep your iPad and its power adapter in a well-ventilated area when in use or charging. Take special care if you have a physical condition that affects your ability to detect heat against the body.

USB power adapter - To operate the Apple USB power adapter safely and reduce the possibility of heat-related injury or damage, plug the power adapter directly into a power outlet. Don't use the power adapter in wet locations, such as near a sink, bathtub, or shower stall, and don't connect or disconnect the power adapter with wet hands. Stop using the power adapter and any cables if any of the following conditions exist:

- The power adapter plug or prongs are damaged.

- The charge cable becomes frayed or otherwise damaged.

- The power adapter is exposed to excessive moisture, or liquid is spilled into the power adapter.

- The power adapter has been dropped, and its enclosure is damaged.

Hearing loss - Listening to sound at high volumes may damage your hearing. Background noise, as well as continued exposure to high volume levels, can make sounds seem quieter than they actually are. Turn on audio playback and check the volume before inserting anything in your ear. For information about how to set a maximum volume limit on iPad.

The Apple headsets sold with iPhone in China (identifiable by dark insulating rings on the plugs) are designed to comply with Chinese standards and are only compatible with iPad, iPhone, and iPod touch.

WARNING: To prevent possible hearing damage, do not listen at high volume levels for long periods.

Radio frequency exposure - iPad uses radio signals to connect to wireless networks. For information about radio frequency (RF) energy resulting from radio signals, and steps you can take to minimize exposure, go to Settings ⚙ > General > About > Legal > RF Exposure.

Radio frequency interference - Observe signs and notices that prohibit or restrict the use of mobile devices. Although iPad is designed, tested, and manufactured to comply with regulations governing radio frequency emissions, such emissions from iPad can negatively affect the operation of other electronic equipment, causing them to malfunction. When use is prohibited, such as while traveling in aircraft, or when asked to do so by authorities, turn off iPad, or use airplane mode or Settings ⚙ > Wi-Fi and Settings > Bluetooth to turn off the iPad wireless transmitters.

Medical device interference - iPad contains components and radios that emit electromagnetic fields. iPad also contains magnets along the left and right edges and back of the device and on the right side of the front glass, which may interfere with medical devices, such as pacemakers and defibrillators. The iPad Smart Cover, iPad Pro Smart Cover, iPad Pro Smart Keyboard, iPad Pro Smart Keyboard Folio, and Apple Pencil (each available separately) also contain magnets. These electromagnetic fields and magnets may interfere with medical devices. Consult your physician and medical device manufacturer for information specific to your medical device and whether you need to maintain a safe distance of separation between your medical device and iPad, the iPad Smart Cover, the iPad Pro Smart Cover, the iPad Pro Smart Keyboard, the iPad Pro Smart Keyboard Folio, and the Apple Pencil. If you suspect iPad is interfering with your medical device, stop using iPad.

Not a medical device iPad is not a medical device and should not be used as a substitute for professional medical judgment. It is not designed or intended for use in the diagnosis of disease or other conditions, or in the cure, mitigation, treatment, or prevention of any condition or disease. Please consult your healthcare provider prior to making any decisions related to your health.

Medical conditions - If you have any medical condition or experience symptoms that you believe could be affected by iPad or flashing lights (for example, seizures, blackouts, eyestrain, or headaches), consult with your physician prior to using iPad.

Explosive and other atmospheric conditions - Charging or using iPad in any area with a potentially explosive atmosphere, such as areas where the air contains high levels of flammable chemicals, vapors, or particles (such as grain, dust, or metal powders), may be hazardous. Exposing iPad to environments having high concentrations of industrial chemicals, including near evaporating liquified gasses such as helium, may damage or impair iPad functionality. Obey all signs and instructions.

Repetitive motion - When you perform repetitive activities such as typing, swiping, or playing games on iPad, you may experience discomfort in your hands, arms, wrists, shoulders, neck, or other parts of your body. If you experience discomfort, stop using iPad and consult a physician.

High-consequence activities - This device is not intended for use where the failure of the device could lead to death, personal injury, or severe environmental damage.

Choking hazard - Some iPad accessories may present a choking hazard to small children. Keep these accessories away from small children.

Important handling information for iPad

Cleaning - Clean iPad immediately if it comes in contact with anything that may cause stains or other damage—for example, dirt or sand, ink, makeup, soap, detergent, acids or acidic foods, and lotions. To clean:

- Disconnect all cables, then do one of the following to turn iPad off:

  - *Models with the Home button:* Press and hold the top button until the slider appears, then drag the slider.

  - *Other models:* Simultaneously press and hold the top buttn and either volume button until the sliders appear, then drag the top slider.

  - *All models:* Go to Settings ⚙ > General > Shut Down, then drag the slider.

    Use a soft, slightly damp, lint-free cloth—for example, a lens cloth.

    Avoid getting moisture in openings.

    Don't use cleaning products or compressed air.

The front of iPad is made of glass with a fingerprint-resistant oleophobic (oil-repellant) coating. This coating wears over time with normal usage. Cleaning products and abrasive materials will further diminish the coating and may scratch the glass.

Using connectors, ports, and buttons - Never force a connector into a port or apply excessive pressure to a button, because this may cause damage that is not covered under the warranty. If the connector and port don't join with reasonable ease, they probably don't match. Check for obstructions and make sure that the connector matches the port and that you have positioned the connector correctly in relation to the port.

Lightning to USB Cable - (for models with a Lightning connector) Discoloration of the Lightning connector after regular use is normal. Dirt, debris, and exposure to moisture may cause discoloration. If your Lightning cable or connector become warm during use or your iPad won't charge or sync, disconnect it from your computer or power adapter and clean the Lightning connector with a soft, dry, lint-free cloth. Do not use liquids or cleaning products when cleaning the Lightning connector.

Lightning to USB Cable or USB-C Charge Cable - (depending on model) Certain usage patterns can contribute to the fraying or breaking of cables. The included cable, like any other metal wire or cable, is subject to becoming weak or brittle if repeatedly bent in the same spot. Aim for gentle curves instead of angles in the cable. Regularly inspect the cable and connector for any kinks, breaks, bends, or other damage. Should you find any such damage, discontinue use of the cable.

Operating temperature - iPad is designed to work in ambient temperatures between 32° and 95° F (0° and 35° C) and stored in temperatures between -4° and 113° F (-20° and 45° C). iPad can be damaged and battery life shortened if stored or operated outside of these temperature ranges. Avoid exposing iPad to dramatic changes in temperature or humidity. When you're using iPad or charging the battery, it is normal for iPad to get warm.

If the interior temperature of iPad exceeds normal operating temperatures (for example, in a hot car or in direct sunlight for extended periods of time), you may experience the following as it attempts to regulate its temperature:

- iPad stops charging.

- The screen dims.

- A temperature warning screen appears.

- Some apps may close.

Important: You may not be able to use iPad while the temperature warning screen is displayed. If iPad can't regulate its internal temperature, it goes into deep sleep mode until it cools. Move iPad to a cooler location out of direct sunlight and wait a few minutes before trying to use iPad again.

Thank you for purchasing our guide. We will keep updating this guide as we discover new tips and tricks and you will get notification anytime there is an update.

Thank you for purchasing our guide.

Index

Z

Printed in Great Britain
by Amazon

61353601R00208